Mathematical Marvels: Texts and Monographs in the Spirit of CR Rao

This series aims to publish books in all areas influenced by Dr. C.R. Rao, including statistics, linear algebra, econometrics and information geometry. It also includes content in other domains of mathematical sciences, applied statistics, financial mathematics, operations research and computer science with primary focus on mathematics and statistics. The series aims to carry on the tradition of C.R. Rao's innovative and influential works, which have impacted a variety of fields, and to provide top-notch textbooks and monographs in related areas.

Arni S. R. Srinivasa Rao

Mathematical Demography: Theory and Modeling

Springer

Arni S. R. Srinivasa Rao
Laboratory for Theory and Mathematical
Modeling, Department of Medicine -
Division of Infectious Diseases
Medical College of Georgia, Augusta
University
Augusta, GA, USA

ISSN 3005-1967 ISSN 3005-1975 (electronic)
Mathematical Marvels: Texts and Monographs in the Spirit of CR Rao
ISBN 978-981-95-6128-5 ISBN 978-981-95-6129-2 (eBook)
https://doi.org/10.1007/978-981-95-6129-2

Mathematics Subject Classification: 91D20, 60K05, 91B30, 35R60, 92D25, 62P10, 62P05, 92D30

This Springer imprint is published by the registered company Springer Nature Singapore Pte Ltd.
The registered company address is: 152 Beach Road, #21-01/04 Gateway East, Singapore 189721, Singapore

Dedicated to Roy M. Anderson, James R. Carey, Jayanta K. Ghosh, Niranjan V. Joshi, Masayuki Kakehashi, Steven G. Krantz, Ronald D. Lee, Philip K. Maini, Robert M. May, Sukumar Mukerji, Vidyanand Nanjundiah, K.B. Pathak, C.R. Rao, J.B. Shukla, Kurapati Sudhakar, and Shripad Tuljapurkar for their influence, inspiration, and enriching interactions over the years.

Preface

The textbook is written clearly and concisely, making the theory and modeling of both deterministic and stochastic easy to understand. Mathematical demography overlaps significantly with various fields, including applied mathematics, probability theory, statistics, actuarial science, epidemiology, and ecology. This book progresses from fundamental concepts to advanced modeling techniques, making it accessible to any interested reader, and provides the necessary theoretical foundations to master the content. Separate chapters in the book discuss classical as well as recent developments in stable and stationary populations, such as Euler-Lotka population equations, and life tables. PDE models of single population dynamics, McKendrick von Foerster models of population growth, and Okubo's diffusion models are derived. Population ecology models like the Lotka-Volterra two-population framework, Kermack-McKendrick three-population models, and their stability analysis foundations are succinctly explained. The book also includes two chapters on stochastic process models in demography and one chapter describing multiple decrement table and survival analysis foundations. A key highlight is its detailed exploration of mathematical demography modeling and theory while maintaining a clear focus on various topics. Several new closed forms and new derivations were done, and newer examples were created exclusively for this book. Several preliminary concepts are mentioned in the footnotes, with additional relevant information found in the Appendix.

In 2024, I began writing a detailed chapter on mathematical demography for an invitation, received from the *Encyclopedia of Social Measurement*, 2nd edition, and a detailed chapter came out. However, there were word limit restrictions for the *Encyclopedia* and I was reluctant to condense the material. Therefore, the chapter

The original version of the book has been revised. A correction to this book can be found at https://doi.org/10.1007/978-981-95-6129-2_9

was withdrawn from there. I wrote an article titled "Mathematical Demography" for Sankhya B (Springer) published in 2025. In fact, the material from my Sankhya B article is taken as is, and several additional materials are added to the book. I am grateful for both these invitations, as they sparked my interest in exploring the topic, which is truly fascinating, and then the idea of writing this book emerged.

There is no definitive scale for determining the appropriate level of technical detail to include; however, I have made every effort to present the content based on my research and teaching experience at various educational levels. I have taught regular academic one-semester courses ranging from undergraduate to Ph.D. programs at institutions such as the Medical College of Georgia at Augusta University in the USA, Indian Statistical Institute in Kolkata, Indian Institute of Science in Bengaluru, and one-week workshops at various other places, including at the International Institute for Population Sciences, Mumbai, India. I am grateful for the excellent research and teaching experiences I received at the institutions mentioned above and several others, including the University of Oxford, Imperial College, Hiroshima University, the University of Guelph, and the Fields Institute in Toronto. I had the opportunity to be affiliated with these institutions through both short-term and long-term academic visits or appointments for about two years or less. The topics I have covered in these courses and workshops include ordinary and partial differential equations, linear algebra, demography, stochastic processes, probability theory, survival analysis, life contingencies, wavelets, mathematical modeling, and mathematical analysis (both real and complex), among others. The content presented here benefits from the tools and techniques used in these courses. The insights gained from collaborations among basic scientists, medical professionals, and engineers were highly beneficial. In addition to my work as an academic, I have applied several mathematical techniques from the book in various government projects, projects from NGOs such as the World Bank, etc., and policies in the capacity of a committee member or as a consultant across multiple countries, particularly assisting in forecasting, estimation, budgeting, and policy formulations. I am grateful for all these experiences, which gave me the opportunity to take mathematical sciences to society.

Throughout the years, I have drawn inspiration from several textbooks that I incorporated into my teaching curriculum. Notably, I would like to highlight the book *Linear Statistical Inference* (1973) by C.R. Rao, which I used for teaching probability theory, and the book *Differential Equations: A Modern Approach with Wavelets* (2021) by S.G. Krantz, which I utilized for teaching courses such as differential equations and wavelets. The style of narration and the direction of these books were very influential. I have been fortunate to collaborate extensively with these two renowned authors, both in editing joint books and in several research collaborations.

The content of the book can be adapted for a one-semester course. The concepts are presented in an engaging, clear, and concise manner for a better reading and learning experience. Significant care has been taken to create all the figures that enhance the learning experience, and the examples provided can be expanded to suit different levels of students. Exercise problems are provided at the end of each chapter. The content of the book is accessible to those interested in learning modeling techniques and grasping the associated theoretical foundations provided in this book.

Augusta, GA, USA
2025

Arni S. R. Srinivasa Rao

Acknowledgements My sincere thanks to Sourabh Bhattacharya, Peeyush Chandra, Joshua Goldstein, Ravindra Khattree, Rajeev Kumar, Tae Jin Lee, Carlos Escudero Liébana, P.D.N. Srinivasu, David Steinsaltz, and Yuehua Wu for taking the time to read either the entire draft or parts of the entire draft of the book carefully and pointing out several suggestions, corrections, and identifying typos, and to Springer's Publishing Editor, Shamim Ahmad, Production Editor, Suresh Dharmalingam, and Production Supervisor, J. Jayasri for their excellent support.

Competing Interests The author has no competing interests to declare that are relevant to the content of this manuscript.

Ethics Approval The author declares that this manuscript does not include primary research with human participants.

The author declares that this manuscript does not include primary studies involving animals.

Contents

List of Figures

List of Tables

Chapter 1
Euler–Lotka Population Integral Equations

Lotka's integral equation, sometimes referred to as Euler–Lotka's integral equation, plays a fundamental role in the formulation of renewal theory. Leonhard Euler worked on population growth models between 1740 and 1760, while Alfred J. Lotka developed his population equations during 1900 to 1945.[1]

Euler–Lotka population equations described in this chapter are among the oldest theories in population dynamics central to the renewal theory of probability, population biology, and ecology modeling. It is also central to other fields of mathematical modeling. These equations are built with four major quantities, namely, $B(t)$ the number of female births at time t for $t \geq 0$, $l(a)$ the survival probability of an individual reaching age a, $m(a)$ the maternity function at age a, and $B(t-a)$ the number of female births in the year $t-a$. Here, a years before t means that $t-a$ also represents a point in time. These quantities will be described further while we build Euler–Lotka equations.

1.1 Renewal Theory of Populations

Before delving into the technicalities of forming the integral equation, it is important to address some preliminaries and provide helpful insights on the quantities required for the process. Further, let $B(0)$ represent the size of a cohort of births (individuals born at the same time), and let $n(a)$ denote the number of survivors at age a from

[1] Euler, among his many contributions, was particularly interested in population analysis during the early stages of his career. He explored questions related to population growth and the estimation of population sizes based on these growth patterns. Euler developed a discrete version of the renewal equation using stationary population approaches, while Lotka created a continuous version of the renewal equation.

A. S. R. Srinivasa Rao, *Mathematical Demography: Theory and Modeling*, Mathematical Marvels: Texts and Monographs in the Spirit of CR Rao,
https://doi.org/10.1007/978-981-95-6129-2_1

the cohort $B(0)$. The survival probability of an individual reaching age a, denoted as $l(a)$, can be computed as follows:

$$l(a) = \frac{n(a)}{B(0)}. \tag{1.1}$$

Here, $n(a)$ for $a \in [0, \omega)$, where ω is the maximum age, is a nonincreasing function.[2] Consequently, the probability function or survival function[3] $l(a)$ is also a nonincreasing function over the age interval $[0, \omega)$. It is crucial to carefully obtain $n(a)$ without considering it with migration data; that is, $n(a)$ must account only for the survivors among $B(0)$. For a continuous age a, we can see that $n(a) \le B(0)$ for $a \ge 0$, and $n(a) \ge n(a + \delta)$ for $\delta > 0$ when $a + \delta \in (0, \omega)$. For any $a \in (0, \omega)$, if $n(a) < B(0)$ holds, then for any $b > a$ (for $b \in (0, \omega)$), it will also hold that $n(b) < B(0)$. Let $D(s)$ represent the number of deaths at age s among $B(0)$ for some $s \in [0, \omega)$. This implies, $B(0) = \int_0^{\omega} D(s)ds$, and we have

$$\int_0^{\omega} \frac{D(s)}{B(0)} ds = 1. \tag{1.2}$$

Let $B(t)$ represent the number of female births at time t for $t \ge 0$. The value of $B(t)$ for $t > 0$ results from the number of childbearing females who were in the reproductive age a years prior to t. Assuming that the reproductive age span for women is 15 to 49 years, it follows that $t - a \ge 15$. For instance, if we consider $B(2025)$ to represent the number of female births in the year 2025, then $B(2025)$ is determined by the number of surviving mothers who were born at least 15 years before 2025 and at most 49 years before 2025. Thus, we can infer that $B(2025)$ is based on surviving mothers who were born between the years 1975 and 2010. We can also express $B(t)$ as the result of surviving females born $t - a$ years earlier, or equivalently, $B(t)$ can be described through the quantity of surviving birth cohort of $B(t - a)$.

The quantity $B(t - 15)$ represents the number of female births in the year $t - 15$. Consequently, $B(t - 15)l(15)$ gives us the number of females[4] who are aged 15 in the year t. The maternity function,[5] at age a denoted as $m(a)$, allows us to calculate the number of new girl babies born to mothers aged a in the year t. Specifically, the

[2] A function $f(a)$ for a values in the interval $[0, \omega)$ is said to be nonincreasing if for all x , $y \in [0, \omega)$ and $x < y$, we have $f(y) \le f(x)$.

[3] The survival function is discussed extensively in the Chapter on Life table. In addition to life tables described in Chap. 3, the survival function of this type, as expressed in (1.1), can also be computed using actual population birth cohorts, denoted as $B(0)$.

[4] The quantity $l(15)$ is the probability of survival to age 15 for the female births in the year $t - 15$. So $B(t - 15)l(15)$ gives the number of females at age 15 in the year t.

[5] The quantity $m(a)$ for the year t can be defined as the rate of female births to mothers of age a in the year t.

number of babies born in the year t to mothers aged 15 can be represented as

$$B(t-15)l(15)m(15). \tag{1.3}$$

For mothers aged 16 in the year t, the number of babies can be computed as $B(t-16)l(16)m(16)$. This pattern continues, allowing us to calculate the number of babies born to mothers aged up to 49 in the year t. Therefore, the birth cohort $B(t)$ for the year t can be derived using the function $l(a)$ in combination with the birth function $B(t-a)$ and the maternity function $m(a)$. The function $m(a)$ represents the probability of a woman aged a giving birth. This leads to the formulation of the basic integral equation of the one-sex number of births in year t, represented by

$$B(t) = \int_{15}^{49} B(t-a)l(a)m(a)da. \tag{1.4}$$

Lotka introduced this integral equation in his famous work on the one-sex population model (see Sharpe and Lotka, 1911 [1]) and in several other articles by Lotka (see, for example, [2–4]). Lotka introduced an additional term, referred to as the initial condition, denoted by $G(t)$. This term quantifies the number of births initially available to women in the reproductive age group at the beginning of the process. The integral equation for $B(t)$ is defined for mothers within the age range of $\alpha = 15$ to $\beta = 49$ is represented by

$$B(t) = \int_{\alpha=15}^{\beta=49} B(t-a)l(a)m(a)da + G(t). \tag{1.5}$$

Equation (1.5) is called the *renewal equation* of populations. Such kind of renewal equations have been studied and analyzed for their properties in the fields of statistics and probability, see, for example, [5–9], and they were also incorporated in textbooks of probability, applied mathematics, ecology, and demography [10–15]. For $t \geq \beta$, we have $G(t) = 0$ since β represents the age at menopause. The births among the women of reproductive age is represented as

$$B(t) = \int_{\alpha}^{\beta} B(t-a)l(a)m(a)da. \tag{1.6}$$

Since $B(t)$ and $B(t-a)$ represent female births occurring at different times, one can consider that the Euler–Lotka equations consist of three major quantities. By substituting $B(t) = Qe^{rt}$ in (1.6) we get

$$\begin{aligned} Qe^{rt} &= \int_{\alpha}^{\beta} Qe^{r(t-a)}l(a)m(a)da, \\ &= \int_{\alpha}^{\beta} Qe^{rt}e^{-ra}l(a)m(a)da. \end{aligned} \tag{1.7}$$

Equation (1.7) is simplified further to obtain

$$\int_{\alpha}^{\beta} e^{-ra} l(a)m(a)da = 1, \tag{1.8}$$

which is recognized as the characteristic equation of Lotka's integral equation. On experimental basis, Lotka assumed $B(t) = Qe^{rt}$. We will see later that this r is the intrinsic rate of natural growth.[6] Prior to that, it was not uncommon to use similar exponential functions as substitutions for solving integral and differential equations. Later in Sect. 1.4, we will examine the solutions of the integral using the substitution of $B(t) = \Sigma_{j=0}^{\infty} Q_j e^{r_j t}$. We will refer to the left-hand side of Lotka's characteristic equation in (1.8) as $\phi(r)$, i.e.,

$$\phi(r) = \int_{\alpha}^{\beta} e^{-ra} l(a)m(a)da, \tag{1.9}$$

where r is called the intrinsic rate of population growth. If $r = 0$, then $\phi(0) = 1$. The quantity $\phi(0)$ is known as the net reproduction rate (NRR) in demography, commonly represented by R_0 (due to Lotka). Therefore, the continuous-age formula for the NRR is expressed as

$$R_0 = \int_{\alpha}^{\beta} l(a)m(a)da. \tag{1.10}$$

When $R_0 = 1$, the female population is expected to replace itself with an average of one female birth throughout her entire reproductive span. The quantity $m(a)$ is interpreted as the probability of a woman giving birth to a female child within the age interval $(a, a + da)$. By using (1.9) and (1.10), we can draw interesting

---

[6] The symbol e is called the natural exponent and e^x for a variable x is called the exponential function, where e^x(or $\exp x$) $= \sum_{x=0}^{\infty} \frac{x^n}{n!}$. Euler's famous formula uses ix instead x for i=$\sqrt{-1}$ in the exponential function, and $e^{ix} = \sum_{x=0}^{\infty} \frac{(ix)^n}{n!} = \cos x + i \sin x$. The famous Laplace transforms of 18th century use exponential functions. Let f be a real-valued function defined on $[0, \infty)$. The Laplace transform of f denoted by $L[f](s)$ is defined [16] as

$$L[f](s) = \int_0^{\infty} \exp(-sx) f(x)dx \ (s > 0).$$

For example, if $f(x) = x$, then its Laplace transform is

$$= \int_0^{\infty} x \exp(-sx)\, dx = -x\frac{1}{s}\exp(-sx)\Big|_0^{\infty} - \int_0^{\infty} \frac{-1}{s}\exp(-sx)$$

$$= 0 + \frac{1}{s}\cdot\left[-\frac{1}{s}\exp(-sx)\right]\Big|_0^{\infty} = \frac{1}{s^2}.$$

Table 1.1 Net reproduction rate (surviving daughters per woman) in various geographic regions. Source: World Population Prospects 2024

	1950–1955	2015–2020
Africa	1.838	1.903
Asia	1.836	0.971
Europe	1.152	0.777
Latin America and the Caribbean	2.144	0.967
Northern America	1.551	0.843
Oceania	1.599	1.085

conclusions about the relationship between R_0 and r. When $R_0 = 1$, it follows that $\phi(0) = 1$. This implies that $\exp(-ra) = 1$, leading to the conclusion that $r = 0$. If $R_0 = 1$, then $r > 0$, and if $R_0 < 0$, then $r < 0$. Although the formula represented in (1.10) theoretically predicts the population replacement level, this interpretation may not hold true in practice due to the stable assumptions surrounding $l(a)$. For further details on these limitations and historical notes on R_0, please refer to Sect. 5.7, Rao (2021) in *PAA Affairs* [17] and Rao (2021) in *IPC 2021 Conference* [18]. Numerous textbooks on technical demography provide additional details and data-based examples on NRR; readers may refer to sources [19–23]. The NRR values for various regions of the world are presented in Table 1.1 based on the data in [24].

1.2 Length of Generation and Birth Rate

Since the birth rate beyond the age interval $[\alpha, \beta]$ within the overall age interval $(0, \infty)$ is zero, one can consider Lotka's characteristic equation as

$$\int_0^\infty e^{-ra} l(a)m(a)da = 1. \tag{1.11}$$

Using Taylor series expansion for e^{-ra} in (1.11) (see Footnote 6), the integral (1.11) becomes

$$\int_0^\infty \left[1 - ra + \frac{r^2a^2}{2!} - \frac{r^3a^3}{3!} + \ldots\right] l(a)m(a)da = 1. \tag{1.12}$$

Retaining only the linear term and omitting higher-order terms

$$\frac{r^n a^n}{n!}$$

for $n = 2, 3, \ldots$ in (1.12), we will get

$$\int_0^\infty [1 - ra]\, l(a)m(a)da = 1.$$

$$\Longrightarrow \int_0^\infty l(a)m(a)da - \int_0^\infty ral(a)m(a)da = 1,$$

$$\Longrightarrow R_0 - r\int_0^\infty al(a)m(a)da = 1,$$

$$\text{(because } R_0 = \int_\alpha^\beta l(a)m(a)da \text{ in (1.10))}$$

$$\Longrightarrow R_0\left(1 - r\frac{\bar{R_0}}{R_0}\right) = 1, \tag{1.13}$$

where $\bar{R_0} = \int_0^\infty al(a)m(a)da$. Taking the logarithm of both sides of (1.13), we obtain

$$\log R_0 + \log\left(1 - r\frac{\bar{R_0}}{R_0}\right) = 0. \tag{1.14}$$

By expanding the term $\log\left(1 - r\frac{\bar{R_0}}{R_0}\right)$ in (1.14) using Taylor series,[7] Eq. (1.14) is written as

$$\log R_0 - r\frac{\bar{R_0}}{R_0} - r^2\frac{\bar{R_0}^2}{R_0^2} - r^3\frac{\bar{R_0}^3}{R_0^3} - \ldots = 0, \tag{1.15}$$

and ignoring higher-order terms beyond the first order, we simplify the expression in (1.15) to express

$$\log R_0 - r\frac{\bar{R_0}}{R_0} = 0. \tag{1.16}$$

Let us denote $\frac{\bar{R_0}}{R_0}$ as L and call L as *the length of a generation* (average age of mothers at the brith of the first baby or the last baby); then it leads us to

$$L \approx \frac{\log R_0}{r}. \tag{1.17}$$

[7] $\log(1 + x) = x - \frac{x^2}{2} + \frac{x^3}{3!} - \frac{x^4}{4} + \ldots$

The NRR or R_0 in (1.10) can be expressed in terms of the intrinsic growth rate (r) and the length of the generation (L) as

$$R_0 = \exp(rL). \tag{1.18}$$

From (1.17), one can write

$$\frac{\bar{R}_0}{R_0} = \frac{\log R_0}{r}.$$

The length of the generation can be defined by approximating r using the first-order Taylor series expansion of $\log\left(1 - r\frac{\bar{R}_0}{R_0}\right)$ as

$$L = \frac{\int_0^\infty a l(a)m(a)da}{\int_0^\infty l(a)m(a)da}. \tag{1.19}$$

The quantity $B(t)$ in (1.6) provides a way to express the total female births at time t. Let $N(a, t)$ represent the size of the population at age a and time t. The total population at time t can be denoted as

$$N(t) = \int_0^\infty N(a, t)da, \tag{1.20}$$

while the birth rate at that time, say, $b(t)$, can be represented by

$$b(t) = \frac{\int_0^\infty B(t-a)l(a)m(a)da}{\int_0^\infty N(a, t)da}. \tag{1.21}$$

Example 1.1 Suppose the net reproduction rate of a country reaches 2.03, and the population is at the stable age distribution[8] with an intrinsic growth rate of 0.03, and then approximate the generation of that population.

Solution The length of generation of the population (L) using (1.17) is computed as

$$L = \frac{\log 2.03}{0.03} \approx 23.6 \text{ years}.$$

[8] A stable age distribution occurs when age-specific fertility and mortality rates remain constant, and the proportion of individuals in each age group stays unchanged over time.

1.3 Intrinsic Birth, Death Rates and Stable Age Distribution

Let $u(a, t)$ be the proportion of the population (with respect to the total population) at age a and at time t. Then the number of individuals of the age group $(a, a + da)$ in the population at time t is $N(t)u(a, t)da$, which can be expressed as

$$N(t)u(a,t)da = B(t-a)l(a)da \text{ (for } a \leq t). \tag{1.22}$$

Integrating (1.22) from 0 to ∞, we get

$$\int_0^{\infty} N(t)u(a,t)da = \int_0^{\infty} B(t-a)l(a)da. \tag{1.23}$$

Since the total of all age fractions in the population sums to 1, i.e., $\int_0^{\infty} u(a,t)da = 1$, we have

$$\int_0^{\infty} N(t)u(a,t)da = N(t)\int_0^{\infty} u(a,t)da = N(t). \tag{1.24}$$

From (1.23) and (1.24), we can express

$$N(t) = \int_0^{\infty} B(t-a)l(a)da. \tag{1.25}$$

Substituting $B(t) = Qe^{rt}$ in (1.25), we get

$$\begin{aligned} N(t) &= Qe^{rt}\int_0^{\infty} e^{-ra}l(a)da \\ &= B(t)\int_0^{\infty} e^{-ra}l(a)da. \end{aligned} \tag{1.26}$$

Using (1.23), (1.25), and (1.26), the number of individuals at age a and at time t is represented by

$$\int_0^{\infty} N(t)u(a,t)da = B(t)\int_0^{\infty} e^{-ra}l(a)da. \tag{1.27}$$

Assuming that birth and death rates are constant over time, we write $b(t) = b$ and $d(t) = d$. Once we obtain a stable birth rate from (1.27), the death rate can be determined as $d = b - r$, where r is the real root of Lotka's fundamental equation in (1.9). Here r is also referred to as the intrinsic rate of natural growth. Using (1.26), we can express

$$b = \frac{B(t)}{N(t)} = \frac{B(t)}{B(t)\int_0^{\infty} e^{-ra}l(a)da} = \frac{1}{\int_0^{\infty} e^{-ra}l(a)da}. \tag{1.28}$$

Under the assumption of a constant age distribution (where $u(a) = \lim_{t\to\infty} u(a,t)$), we have $\int_0^\infty u(a)da = \int_0^\infty u(a,t)da = 1$, and

$$u(a,t) = \frac{N(a,t)}{N(t)} = \frac{B(t-a)l(a)}{\int_0^\infty B(t-a)l(a)da}. \tag{1.29}$$

In (1.29), we acknowledge that after adjusting for deaths, the total population $N(t)$ is equal to $\int_0^\infty B(t-a)l(a)da$. By substituting $B(t) = Qe^{rt}$, we ultimately derive the stable age distribution as

$$u(a) = \frac{e^{-ra}l(a)}{\int_0^\infty e^{-ra}l(a)da} = be^{-ra}l(a). \tag{1.30}$$

Example 1.2 Suppose we want to compute the stable birth rate b and death rate d of a country or a region. If the life expectancy at birth is 75 years, then the birth rate b can be calculated as $\frac{1}{75} = 13.3$ (per 1000 population). If we know that $r = 0.0067$, we can determine the stable death rate d using the equation:

$$d = b - r = 13.3 - 6.7 = 6.6 \text{ (per 1000 population)}.$$

Example 1.3 For a stable population similar to Example 1.2, let us write d in terms of b and r using b in (1.28) as follows:

$$d = b - r = \frac{1}{\int_0^\infty e^{-ra}l(a)da} - r. \tag{1.31}$$

Let $D = \frac{1}{\int_0^\infty e^{-ra}l(a)da} - r$. We will find $\frac{dD}{dr}$ and solve for r by making $\frac{dD}{dr} = 0$. We have

$$\frac{dD}{dr} = \frac{\int_0^\infty ae^{-ra}l(a)da}{\left(\int_0^\infty e^{-ra}l(a)da\right)^2} - 1.$$

Assuming the change in the deaths with respect to the intrinsic growth rate is zero, we get

$$\frac{\int_0^\infty ae^{-ra}l(a)da}{\left(\int_0^\infty e^{-ra}l(a)da\right)^2} - 1 = 0 \implies \frac{\int_0^\infty ae^{-ra}l(a)da}{\left(\int_0^\infty e^{-ra}l(a)da\right)^2} = 1$$

$$\implies \frac{\int_0^\infty ae^{-ra}l(a)da}{\int_0^\infty e^{-ra}l(a)da} \cdot \frac{1}{\int_0^\infty e^{-ra}l(a)da} = 1. \tag{1.32}$$

The first term on the left side of (1.32) is the mean age of the stable population, and the second term is the stable birth rate.

We can also express

$$\frac{db}{dr} = \bar{a}b, \tag{1.33}$$

where $\bar{a}$ is the mean age of the stable population and is expressed using Example 1.3 by

$$\bar{a} = \frac{\int_0^\infty ae^{-ra}l(a)da}{\int_0^\infty e^{-ra}l(a)da}. \tag{1.34}$$

It is evident that r and $\bar{a}$ are inversely related. The change in stable age distribution with respect to intrinsic rate of nature growth is

$$\begin{aligned}
\frac{du(a)}{dr} &= \frac{d}{dr}\left[be^{-ra}l(a)\right] \\
&= \frac{db}{dr}\left(e^{-ra}l(a)\right) - bae^{-ra}l(a) \\
&= bae^{-ra}l(a)\left(\frac{1}{b}\frac{db}{dr} - a\right) \\
&= u(a)\left(\bar{a} - a\right)\left(\text{since}\frac{db}{dr} = \bar{a}b\text{from (1.33)}\right).
\end{aligned} \tag{1.35}$$

For $a = \bar{a}$, there will be no change in the stable age distribution, i.e. for $a = \bar{a}$, we will have $\frac{du(a)}{dr} = 0$. From (1.35), it is evident that $u(a)$ and r are linearly related, and we saw above that r and $\bar{a}$ are inversely related. Therefore, for higher r values in a population, there will be a higher proportion of younger people. Also, an increase in r decreases $\bar{a}$.

Example 1.4 Using (1.9), let $\phi(r) = \int_\alpha^\beta e^{-ra}l(a)m(a)da$. Show that

$$\phi(r) = R_0 e^{-\int_0^r Xrdr}, \tag{1.36}$$

where

$$X = \frac{\int_\alpha^\beta ae^{-ra}l(a)m(a)da}{\int_\alpha^\beta e^{-ra}l(a)m(a)da}. \tag{1.37}$$

Proof We note that

$$\frac{d\phi(r)}{dr} = -\int_\alpha^\beta ae^{-ra}l(a)m(a)da,$$

From the given expression of X in (1.37), we can write

$$\frac{d\phi(r)}{dr} = -X \int_{\alpha}^{\beta} e^{-ra} l(a)m(a)da$$
$$\implies \frac{d\phi(r)}{dr} = -X\phi(r). \tag{1.38}$$

We separate the variables in (1.38), and integrate with respect to r, to get

$$\log \phi(s) = -\ Xs|_0^r + C \text{ (for a constant of integration } C\text{)},$$
$$\implies \phi(r) = e^{-Xr} e^C.$$

When $r = 0$, it becomes $\phi(0) = e^C$ and, as we referred to previously, $\phi(0) = R_0$. So, we arrive at the exponential model in (1.36).[9] □

1.4 Real and Complex Roots of Renewal Equation

Let

$$B(t) = \Sigma_{j=0}^{\infty} Q_j e^{r_j t}$$

be a solution of the renewal equation

$$B(t) = \int_0^{\infty} B(t-a)l(a)m(a)da. \tag{1.39}$$

This implies

$$\begin{aligned} \sum_{j=0}^{\infty} Q_j e^{r_j t} &= \int_0^{\infty} \sum_{j=0}^{\infty} Q_j e^{r_j(t-a)} l(a)m(a)da, \\ &= \int_0^{\infty} \sum_{j=0}^{\infty} Q_j e^{r_j t} e^{-r_j a} l(a)m(a)da, \\ &= \sum_{j=0}^{\infty} Q_j e^{r_j t} \int_0^{\infty} e^{-r_j a} l(a)m(a)da. \end{aligned} \tag{1.40}$$

[9] The population model $\frac{dN}{dt} = rN$ for the growth rate r can be written as $\int \frac{dN}{N} = r \int dt, \implies \ln N(t) = rt + C$ for a constant of integration C. This is also referred to as the Malthusian population model. Thus, $N(t) = e^{rt}e^C = N(0)e^{rt}$, where $N(0) = e^C$.

Simplifying (1.40), we obtain

$$\int_0^\infty e^{-r_j a} l(a)m(a)da = 1 \text{ for } j = 0, 1, 2, \ldots. \tag{1.41}$$

Let us denote the L.H.S. of the above integral (1.41) as $\phi(r_j)$. We note below two things:

$$\frac{d\phi(r_j)}{dr_j} = -\int_0^\infty a e^{-r_j a} l(a)m(a)da < 0,$$

because $a \geq 0$, and $l(a)m(a) > 0$, and

$$\frac{d^2\phi(r_j)}{dr_j^2} = \int_0^\infty a^2 e^{-r_j a} l(a)m(a)da > 0.$$

This concludes that $\phi(r_j)$ is a strictly decreasing continuous function of r_j. Since $\phi(r_j)$ is strictly decreasing, it intersects the line $\phi(r_j) = 1$ exactly once, indicating that the function has one real solution and infinitely many complex number solutions. See Fig. 1.1. For further mathematical details on complex numbers and their conjugates, see Appendix C. Let r_0 be the one real root of $\phi(r_j) = 1$ and r_j for $j = 1, 2, \ldots$ represent the complex roots of $\phi(r_j) = 1$. This ensures that

$$\int_0^\infty e^{-r_0 a} l(a)m(a)da = 1, \tag{1.42}$$

where r_0 is the real root of the characteristic equation. We can express complex numbers as $r_j = a_j + ib_j$ for $j = 1, 2, 3, \ldots$, where a_j is the real part and b_j

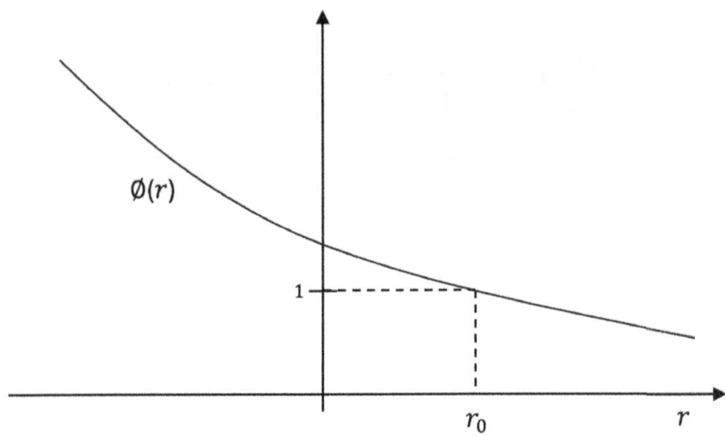

Fig. 1.1 The value of one real root of $\phi(r) = 1$ can be determined using $\phi(r)$ curve

is the imaginary part. Substituting $r_j = a_j + ib_j$ into the characteristic equation $\phi(r_j) = 1$ transforms it into

$$\int_0^\infty e^{-(a_j+ib_j)a} l(a)m(a)da = \int_0^\infty e^{-a_j a} e^{-ib_j a} l(a)m(a)da = 1. \tag{1.43}$$

Using $e^{i\theta} = \cos\theta + i\sin\theta$ in (1.43), we get

$$\int_0^\infty e^{-a_j a} \left[\cos(-b_j a) + i\sin(-b_j a)\right] l(a)m(a)da = 1\ (1 = 1 + 0i). \tag{1.44}$$

Equating real and imaginary parts of (1.44), we get

$$\int_0^\infty e^{-a_j a} \cos(-b_j a) l(a)m(a)da = 1,$$

$$\int_0^\infty e^{-a_j a} \sin(-b_j a) l(a)m(a)da = 0.$$

For a given value of b_j, we see that

$$\int_0^\infty e^{-a_j a} \cos(-b_j a) l(a)m(a)da < \int_0^\infty e^{-a_j a} l(a)m(a)da, \tag{1.45}$$

and we have already seen that

$$\int_0^\infty e^{-r_0 a} l(a)m(a)da = 1. \tag{1.46}$$

This gives us

$$\int_0^\infty e^{-r_0 a} l(a)m(a)da < \int_0^\infty e^{-a_j a} l(a)m(a)da \text{ (for the real } a_j). \tag{1.47}$$

From (1.45)–(1.47), we can conclude that the value of r_0 is greater than the real part of r_j for $j = 1, 2, 3, \ldots$, i.e., $r_0 > a_j$. The solution of the renewal equation is represented as

$$B(t) = Q_0 e^{r_0 t} + \sum_{j=1}^\infty Q_j e^{r_j t}. \tag{1.48}$$

In (1.48), r_0 is called an intrinsic growth rate or intrinsic rate of natural variation. In a stable population, the size of the population $N(t)$ is calculated by dividing the number of births $B(t)$ by the birth rate of that population. The structure of the equations with complex roots generates transient behavior seen in the model.

Example 1.5 Suppose for the stationary population described in Example 1.2 there are 506,643 number of births occurred. Then the total population $N(t)$ is computed as

$$N(t) = \frac{506{,}643 \times 1000}{13.1} = 38{,}675{,}038.$$

Next, we will show a theoretical example of the relationship between the function $u(a)$ and mean age of the stable population.

Example 1.6 Show that $\frac{du(r,a_0)}{dr} = be^{-ra_0}l(a_0)[\bar{a} - a_0]$, where $u(r, a_0)$ the function $u(a)$ at a specific age a_0, and $\bar{a}$ is expressed in (1.34).

Solution The function $u(r, a_0)$ can be represented as

$$u(r, a_0) = \frac{e^{-ra_0}l(a_0)}{\int_0^\infty e^{-ra}l(a)da}.$$

Taking the derivative of $u(r, a_0)$ with respect to r, we get

$$\begin{aligned} \frac{du(r, a_0)}{dr} &= \frac{-ae^{-ra_0}l(a_0) \int_0^\infty e^{-ra}l(a)da + e^{-ra_0}l(a_0) \int_0^\infty ae^{-ra}l(a)da}{\left(\int_0^\infty e^{-ra}l(a)da\right)^2} \\ &= be^{-ra_0}l(a_0)[\bar{a} - a_0], \end{aligned}$$

where b is the stable birth rate or intrinsic birth rate.

The subject of mathematical demography has been influential in developing important ideas of practical and theoretical developments in understanding populations for the past several centuries. Building models to understand the population dynamics by different age group models for population growth and predicting longevity have been interesting since the medieval days. Lotka's integral equations also played fundamental role in deriving a formula for the NRR. For more references on stable population theory and their applications, refer to [9, 25–30]. Later in Chap. 2, we will see that applications of such integral equations arise in the analysis of McKendrick von Foerster type of partial differential equations. In Sect. 5.7, we will study the limitations of interpreting the computed values of NRR in nonstable populations.

1.5 Exercises

Exercise 1.7 In a stable population, the intrinsic growth rate is given as $r = 0.8$ and maximum age $\omega = 105$. If $l(a) = e^{-ra^2}$, then find the stable birth rate of this population.

Exercise 1.8 Suppose the net reproduction rate of a country reaches 1.96, and the population is at the stable age distribution with an intrinsic growth rate of 0.04. Then by ignoring higher-order Taylor expansion terms, evaluate the integral $\int_0^\infty al(a)m(a)da$.

Exercise 1.9 Let $\bar{R_0} = \int_0^\infty al(a)m(a)da$. Ignoring higher-order terms of $\left(\frac{\bar{R_0}}{R_0}\right)^n$ for $n \geq 3$, show that

$$r = -\frac{1}{2}\left(\frac{R_0}{\bar{R_0}}\right) \pm \frac{\sqrt{\left(\frac{\bar{R_0}}{R_0}\right)^2 (1 + 4\log R_0)}}{2\left(\frac{\bar{R_0}}{R_0}\right)^2}.$$

Exercise 1.10 Let $r_0 = r_1 + k$ in Lotka's characteristic equation. When k is small, show that

$$k = \frac{\int_0^\infty e^{-r_1 a} l(a)m(a)da - \int_0^\infty e^{-r_0 a} l(a)m(a)da}{\int_0^\infty ae^{-r_0 a} l(a)m(a)da}.$$

Hint. Since k is small, use $e^{-ka} \approx (1 - ka)$.

Exercise 1.11 Given a country's population in the year 2000 was 437,000, and it was 682,000 in 2025, suppose the projected population for the same country is 961,500. Build a logistic growth model for the population $P(t)$ using the data provided. ***Hint.*** Use

$$682{,}000 = \frac{437{,}000a}{437{,}000 + \left(a - 437{,}000b\right)e^{-25a}},$$

and

$$961{,}500 = \frac{437{,}000a}{437{,}000 + \left(a - 437{,}000b\right)e^{-50a}},$$

and then compute a and b.

Exercise 1.12 Let $\mu(a)$ be the instantaneous death rate or force of mortality at age a. The total female deaths at time t are represented by

$$\int_0^\infty B(t-a)l(a)\mu(a)da,$$

and the number of females aged a at time t is represented by $B(t-a)l(a)$. The crude death rate, say, D, is defined by

$$D = \frac{\int_0^\infty B(t-a)l(a)\mu(a)da}{\int_0^\infty B(t-a)l(a)da}.$$

Show that

$$D = \frac{1}{\int_0^\infty B(t-a)l(a)da} - r.$$

Hint. Use the number of births $B(t) = \sum_{i=1}^{\infty} Q_i e^{r_i t}$, where $r_1, r_2, \ldots$ are the roots.

References

1. Sharpe, F.R., Lotka, A.J.: A problem in age-distribution. Philos. Mag. **21**, 435–438 (1911)
2. Lotka, A.J.: The stability of the normal age distribution. Proc. Natl. Acad. Sci. **8**(11), 339–345 (1922)
3. Lotka, A.J.: Some recent results in population analysis. J. Am. Stat. Assoc. **33**(201), 164–178 (1938)
4. Lotka, A.J.: A contribution to the theory of self-renewing aggregates, with special reference to industrial replacement. Ann. Math. Stat. **10**(1), 1–25 (1939)
5. Feller, W.: On the integral equation of renewal theory. Ann. Math. Stat. **12**, 243–267 (1941)
6. Karlin, S.: On the renewal equation. Pacific J. Math. **5**, 229–257 (1955)
7. Hadeler, K.P.: Integral equations for infections with discrete parasites: hosts with Lotka birth law. Mathematical Ecology (Trieste, 1982), pp. 356–365. Lecture Notes in Biomath., vol. 54. Springer, Berlin (1984)
8. Deligönül, Z.: Şeyda An approximate solution of the integral equation of renewal theory. J. Appl. Probab. **22**(4), 926–931 (1985)
9. United Nations:. The Concept of a Stable Population. United Nations, New York (1969)
10. Feller, W.: An Introduction to Probability Theory and Its Applications, vol. I, xviii+509 pp., 3rd edn. Wiley, New York, London, Sydney (1968).
11. Smith, D.P., Keyfitz, N.: Mathematical Demography. Selected papers. Second, revised edition. Edited and with commentaries by Kenneth W. Wachter and Hervé Le Bras. Demographic Research Monographs, xxiv+335 pp. Springer, Heidelberg, (2013). ISBN: 978-3-642-35857-9; 978-3-642-35858-6
12. Keyfitz, N., Caswell, H.: Applied Mathematical Demography, 3rd edn, xxi+441 pp. Springer Texts in Statistics. Springer, New York (2013)
13. Rao, A.S.R.S., Bai Z.D., Rao, C.R. (eds.): Probability Models, Handbook of Statistics, vol. 51. Elsevier (2024)
14. Hoppensteadt, F.: Mathematical Theories of Populations: Demographics, Genetics and Epidemics. Society for Industrial and Applied Mathematics, Philadelphia, PA (1975)
15. Kot, M.: Elements of Mathematical Ecology, x+453 pp. Cambridge University Press, Cambridge (2001)
16. Krantz, S.: Differential Equations: A Modern Approach with Wavelets. Chapman and Hall/CRC (2020)

17. Rao, A.S.R.S.: Is NRR Time-Sensitive in Measuring Population Replacement Level. PAA Affairs. PAA: Population Association of America (2021). https://www.populationassociation.org/blogs/emily-merchant1/2021/01/26/is-nrr-time-sensitive
18. Rao, A.S.R.S.: Is NRR Time-Sensitive in Measuring Population Replacement Level, Presented in session 83. Data and methods: A medley of perspectives. In: International Population Conference, Hyderabad, India (2021), IPC 2021. https://ipc2021.popconf.org/abstracts/210503
19. Carey, J.R., Roach, D.: Biodemography. An Introduction to Concepts and Methods. Princeton University Press, Princeton (2019)
20. Ramkumar, R.: Technical Demography. New Age International Press, New Delhi (1986)
21. Pathak, K.B., Ram, F.: Techniques of Demographic Analysis. Himalaya Publishing House, New Delhi (1992)
22. Preston, S., Heuveline, P., Guillot, M.: Demography: Measuring and Modeling Population Processes, 1st edn. Wiley, Hoboken (2000)
23. Misra, B.D.: An Introduction to the Study of Population. South Asian Publishers, New Delhi (1982)
24. United Nations: United Nations, Department of Economic and Social Affairs, Population Division (2024). World Population Prospects 2024: Data Sources. (UN DESA/POP/2024)
25. Rao, A.S.R.S., Carey, J.R.: Stationary status of discrete and continuous age-structured population models. Math. Biosci. **364**, 109058 (2023). Elsevier
26. Shryock, H., Siegel, J., Stockwell, E.: The Methods and Materials of Demography (Condensed Edition). Academic Press, New York (1976)
27. Mitra, S.: Influence of instantaneous fertility decline to replacement level on population growth. Demography **13**, 513–519 (1976)
28. Cerone, P.: The long-term effects of time-dependent maternity behavior. Demography **20**(1), 79–86 (1983)
29. Cerone, P., Keane, A.: The Stable Births Resulting from a time dependent change between two net maternity functions. Demography **15**(1), 135–137 (1978)
30. Mitra, S.: Generalization of the immigration and the stable population model. Demography **20**, 111–115 (1983)

Chapter 2
PDE Models in Population Dynamics

The necessity to build a PDE (partial differential equation) model in demography arises when we need to understand the influence of the rate of change of more than one variable in the population growth. The dynamics of a population are considered here with respect to age and time.[1] Here age and time are considered as continuous. One could include other kinds of rates of change, such as influence on population growth due to changes in age, time, environment, food, etc. In this chapter, a partial differential equation model for the population with age structure is derived. Such models were initially considered by McKendrick and later by von Foerster. Further, a PDE model for population aging developed by Samuel Preston and others is presented. Relaxing certain assumptions of their model, a newer form of the aging model to understand the dynamics of aging, is developed later in this chapter. Finally, a diffusion model proposed by Akira Okubo is derived.

2.1 First-Order PDE Population Model

Let us consider a first-order PDE of the form in (2.1) with two variables, namely, $n = n(a, t)$ representing population density of age a at time t and $\mu = \mu(a)$ representing the *force of mortality* at age a, respectively:

$$\frac{\partial n}{\partial a} + \frac{\partial n}{\partial t} + \mu(a)n = 0. \tag{2.1}$$

[1] Let N be the size of the population at time t. When the dynamics of N is considered with respect to only time t, then $\frac{dN}{dt} = rN$ (due to Malthus) is an ordinary differential equation (ODE) with a growth rate r. This will lead to an exponential growth model. Let N be a function of both age and time. One could consider the dynamics of N with respect to age a and time t by introducing a symbol ∂ (called partial). The expressions $\frac{\partial N}{\partial t}$ and $\frac{\partial N}{\partial a}$ are partial derivatives.

A. S. R. Srinivasa Rao, *Mathematical Demography: Theory and Modeling*,
Mathematical Marvels: Texts and Monographs in the Spirit of CR Rao,
https://doi.org/10.1007/978-981-95-6129-2_2

One can consider $n(a,t)$ in (2.1) as the size of the population of age a at time t in a geographical area, and $\mu(a) = \lim_{\Delta t \to 0} \frac{l(a)-l(a+\Delta t)}{\Delta t l(a)}$. Refer to Chap. 3 (Sect. 3.1) for further details on the force of mortality function. In the current context it will be $n(t) = \int_0^\infty n(a,t)da$. In Chap. 3 (Sect. 3.3) it is shown that

$$l(a) = l(0)\exp\left(-\int_0^a \mu(s)ds\right), \tag{2.2}$$

where $l(a)$ is the number of survivors of the initial population cohort of $l(0)$ individuals. See Chap. 1 for definitions of $l(a)$ and $l(0)$. The population model (2.1) as defined on *age-time* plane, is shown in Fig. 2.1. The behavior of the model at the initial time $t = 0$ is described by the conditions $n(a,0) = n_0(a)$ for $a \geq 0$. Here $n_0(a)$ represents the initial age distribution of the population. Further, the population $n(a,t)$ should satisfy the birth law for $t > 0$ which is given by

$$n(0,t) = \int_0^\infty n(a,t)m(a)da, \tag{2.3}$$

where $m(a)$ is the maternity function of the female population which is at an age a, or probability of a woman aged a giving birth. If we represent n parametrically at

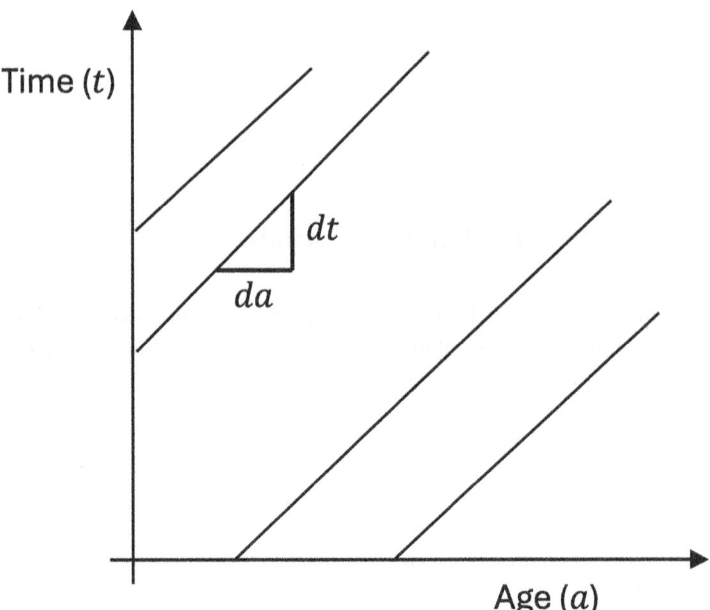

Fig. 2.1 Lexis diagram representation of McKendrick population model

$a = a(t^*)$ and $t = t(t^*)$ for a fixed t^* such that $n^*(t^*) = n\left(a(t^*), t(t^*)\right)$, then

$$\frac{dn^*(t^*)}{dt^*} = \frac{\partial n}{\partial a}\frac{da}{dt^*} + \frac{\partial n}{\partial t}\frac{dt}{dt^*}. \tag{2.4}$$

The method of characteristics is a technique for solving PDEs, and such methods are used for solving the PDE in (2.1). We outline such methods later in this chapter. Taking $a = a(0) + t^*$ and $t = t(0) + t^*$ and treating $n^*(0)$ as initial population at a point $(a(0), t(0)) = (a_0, t_0)$, the population density $n(a, t)$ along the curve $(a(t^*), t(t^*))$ given by $n^*(t^*)$ can be determined using (2.2), and it is represented by

$$n^*(t^*) = n^*(0)\exp\left(-\int_0^{t^*} \mu(a_0 + s)ds\right). \tag{2.5}$$

2.2 McKendrick–von Foerster Equations

Let $n(a, t)$ be the number of individuals in the age group $(a, a + da)$ at time t. Here $a + da$ is treated as a small incremental to a. Since $n(a, t)$ is considered as the number within a continuous age range, one could treat $n(a, t)$ as a population density at age a and time t. Changes in the compartment of $n(a, t)$ occur due to the entry of new individuals who become aged a, individuals who are leaving $n(a, t)$ due to aging, and individuals leaving $n(a, t)$ compartment due to deaths and migration. Aging occurs due to the change in time, so both age and time are changing in the population of $n(a, t)$. Let $N(t)$ be the total population at time t (population of all ages). The quantity $N(t)$ can be represented by

$$N(t) = \int_0^{\infty} n(a, t)da. \tag{2.6}$$

During the time interval $(t, t+\delta t)$, the population $n(a, t)$ also ages $a+\delta t$ units older. That is, without any deaths in $n(a, t)$ during $(t, t + \delta t)$, the change in population of $n(a, t)$ occurs only due to the aging factor (by ignoring migration). If we consider the number of deaths during $(t, t + \delta t)$ at a rate of death $d(a, t)$ (death rate $d(a, t)$ at age a and time t), then we can express

$$n(a + \delta t, t + \delta t) - n(a, t) = -d(a, t)n(a, t)\delta t \tag{2.7}$$

$$\Longrightarrow \quad \frac{n(a + \delta t, t + \delta t) - n(a, t)}{\delta t} = -d(a, t)n(a, t). \tag{2.8}$$

By taking the limit as $\delta t \to 0$, we get

$$\lim_{\delta t \to 0} \frac{n(a+\delta t, t+\delta t) - n(a,t)}{\delta t} = -\lim_{\delta t \to 0} d(a,t)n(a,t)$$

$$\frac{\partial n(a,t)}{\partial t} + \frac{\partial n(a,t)}{\partial a} = -d(a,t)n(a,t). \tag{2.9}$$

The PDE model in (2.9) is of first order.[2] This equation can be rewritten using another notation for PDEs as

$$n_t(a,t) + n_a(a,t) + d(a,t)n(a,t) = 0 \quad (0 < a < \infty, t > 0). \tag{2.10}$$

The equation in (2.10) can also be expressed as

$$n_t + n_a + dn = 0 \quad (0 < a < \infty, t > 0). \tag{2.11}$$

The notations n_t and n_a in (2.11) represent $\frac{\partial n(a,t)}{\partial t}$ and $\frac{\partial n(a,t)}{\partial a}$, respectively. Let us now introduce a birth rate $b(a,t)$ at age a and time t on the population $n(a,t)$. We can see that the total population at t is represented as

$$N(t) = \int_0^\infty n(a,t)da,$$

and the total population at $t = 0$ is represented as

$$N(0) = \int_0^\infty n(a,0)da. \tag{2.12}$$

During the time interval $(t, t+\delta t)$, the quantity $N(0)$ would grow to

$$N(0)\delta t = \delta t \int_0^\infty n(a,0)da. \tag{2.13}$$

The number of births given by $n(a,t)$ would be

$$b(a,t)n(a,t). \tag{2.14}$$

[2] One can derive Eq. (2.8) alternatively as follows: Let $n(a,t)\delta a$ be the number of individuals at time t who are in the age group $(a, a+\delta a)$. Let us assume that the change in the population is caused only by deaths. Let $d(a,t)$ be the death rate at age a and at time t. The number of deaths during $(t, t+\delta t)$ among $n(a,t)\delta a$ can be expressed by $n(a,t)\delta a d(a,t)\delta t$. This gives us

$$n(a+\delta t, t+\delta t)\delta a - n(a,t)\delta a = -d(a,t)\delta a n(a,t)\delta t,$$

and we arrive at (2.8).

The population size during $(t, t+\delta t)$ changed due to the total births during $(t, t+\delta t)$ which can be expressed as

$$N(t)\delta t \int_0^\infty b(a,t)n(a,t)da. \tag{2.15}$$

The birth rate during $(t, t+\delta t)$ using (2.15) would be

$$b(t) = \frac{N(t)\delta t \int_0^\infty b(a,t)n(a,t)da}{N(t)\delta t} = \int_0^\infty b(a,t)n(a,t)da. \tag{2.16}$$

We will use characteristic functions to express $n(a,t)$ and solve $b(t)$ in (2.16) as follows:

Substituting

$$n(a,t) = \begin{cases} B(t-a)\exp\left\{-\int_0^a d(s, t-a+s)ds\right\} & \text{for } a < t \\ n_0(a-t)\exp\left\{-\int_0^t d(a-t+s, s)ds\right\} & \text{for } a \geq t \end{cases} \tag{2.17}$$

in (2.16), we get

$$\begin{aligned} b(t) &= \int_0^t b(a,t)n(a,t)da + \int_t^\infty b(a,t)n(a,t)da \\ &= \int_0^t b(a,t)B(t-a)e^{-\int_0^a d(s,t-a+s)ds}da \\ &+ \int_t^\infty b(a,t)n_0(a-t)e^{-\int_0^t d(a-t+s,s)ds}da \\ &= G(t) + \int_0^t b(a,t)B(t-a)e^{-\int_0^a d(s,t-a+s)ds}da, \end{aligned} \tag{2.18}$$

where $n_0(a) = n(a,0)$ is the initial age distribution of the population, and $G(t) = \int_t^\infty b(a,t)n_0(a-t)\exp\left\{-\int_0^t d(a-t+s,s)ds\right\}da$. We assume the initial age distribution of the population is known, say $n_0(a) = n(a,0)$. The function $B(t)$ can be further modified to express

$$b(t) = G(t) + \int_0^t G(a,t)B(t-a)da, \tag{2.19}$$

where $G(a,t) = b(a,t)e^{-\int_0^a d(s,t-a+s)ds}$. The value of $G(t)$ becomes *zero* for $a \geq t$ because of completion of reproduction age of the female population. This implies $B(t)$ can be expressed as

$$B(t) = \int_0^t G(a,t)B(t-a)da. \tag{2.20}$$

The birth rate in (2.20) resembles the form of Euler–Lotka population renewal equations that we came across in Chap. 1. The above analysis indicates the population renewal theory and McKendrick von Foerster's PDE equation analysis are associated in understanding the population growth. Such types of renewal equations are also applied frequently in epidemic analysis, for example, refer to [1]. The McKendrick von Foerster model can be summarized as

$$\begin{aligned} n_t(a,t) + n_a(a,t) + d(a,t)n(a,t) = 0 \quad (0 < a < \infty, t > 0) \\ b(t) = \int_0^\infty b(a,t)n(a,t)da \quad (t > 0) \\ n_0(a) = n(a,0) \quad (0 < a < \infty). \end{aligned} \tag{2.21}$$

Using $N(t)$, $b(t)$, and $d(t)$ obtained through the PDE framework explained above, we express the change in $N(t)$ over the time as an ordinary differential equation:

$$\frac{dN(t)}{dt} = [b(t) - d(t)]\, N(t). \tag{2.22}$$

The solution for $N(t)$ from (2.22) is obtained as

$$N(t) = N(0)e^{\int_0^t [b(s)-d(s)]ds}, \tag{2.23}$$

where $N(0)$ is also derived out of constant of integration. The quantity $b(t) - d(t)$ in (2.22) is called Malthusian parameter or natural growth rate of the population. If $b(t) > d(t)$, the population $N(t)$ grows, and if $b(t) < d(t)$, the population $N(t)$ decays, and if $b(t) = d(t)$, then the population attains *zero population growth*. For further reading on McKendrick von Foerster equations, refer to [2–14]. In other chapters (Chap. 3 and Chap. 5), we will introduce zero growth models like the life table models where the population size is constant and obey other properties of stationary population models. The characteristic equation of the type used in (2.17) is frequently applied in the numerical analysis of McKendrick von Foerster equations and other such first-order PDEs. The characteristic equation approach is one of the regular methods of solving PDEs. See, for example, [8, 9, 15–17]. Refer to further texts listed above for more details on such PDE models in population dynamics. In Chap. 4, we study characteristic roots while understanding the stability of a system of ODEs.

2.3 Aging Model

In this section, we introduce a general PDE model structured to understand the aging process in a population. This model is developed from the foundational principles of population dynamics and builds upon the earlier model proposed by Preston, Himes, and Eggers (1989) [18]. The original aging model aimed to analyze the rates of aging and the mean age of the population. In this context, birth and death rates play essential roles in understanding the rate of aging of a population at time t and the corresponding change in its mean age.

However, the original model did not consider the birth rates within the interval $[t, t + \Delta t]$ for the population that was alive at time t. To address this limitation, we present a revised model that incorporates these factors and includes suggested improvements. This updated model can be viewed as a preliminary version before conducting a more comprehensive analysis and data fitting. Furthermore, the new model is theoretically more flexible in assessing the population's aging status.

Let us consider the following population aging model:

$$\frac{dA(t)}{dt} = A_P \left[\frac{\partial N(t)}{\partial t}\right] + N(t) \left[\frac{\partial A_P(t)}{\partial t}\right], \tag{2.24}$$

where $A(t)$ is the sum of the ages of all the population who are alive at t, and $N(t)$ is that population alive at t. The quantity $A_p(t)$ is the average age of $N(t)$. Equation (2.24) was obtained from the relation $A(t) = N(t)A_P(t)$. It is a consequence of Leibniz rule of derivation of a product. See [18]. The above model was obtained by assuming the rate of change in the age added in the interval $[t, t+\Delta t]$ due to the births in the interval being zero. Next, we relax this assumption to derive an expression for the population aging and add boundary conditions to the model.

2.3.1 *Relaxing the Assumption on Birth Rates*

Let us study changes in $A(t)$ during the interval $[t, t+\Delta t]$ for $\Delta t > 0$. The changes in $A(t)$ during $[t, t + \Delta t]$ occur due to the following three events:

(i) Total births during $[t, t+\Delta t]$ who survive through the interval and the number of births who died during the interval
(ii) Total deaths among $N(t)$ during $[t, t + \Delta t]$
(iii) All the survivors of $N(t)$ during $[t, t + \Delta t]$

The quantity $A(t)\Delta t$ indicates the total age of the population at $t+\Delta t$ if no deaths during $[t, t+\Delta t]$. We are initially interested in finding the quantity of aged person-years during $[t, t+\Delta t]$, which is

$$N(t)\Delta t - \left\{\begin{array}{c}\text{the total age lost due to}\\ \text{deaths during } [t, t+\Delta t]\end{array}\right\} + \tag{2.25}$$

$$\left\{\begin{array}{c}\text{the total age gained due to}\\ \text{births during } [t, t+\Delta t] \text{ who did}\\ \text{not die during the interval}\end{array}\right\}.$$

If no event of births and deaths occur, then

$$A(t+\Delta t) = A(t) + N(t)\Delta t, \tag{2.26}$$

$\Longrightarrow$

$$\lim_{\Delta t \to 0} \frac{A(t+\Delta t) - A(t)}{\Delta t} = N(t). \tag{2.27}$$

Let us consider the quantity $A(t, t+\Delta t) - A(t)$ when the births and deaths occur during $[t, t+\Delta t]$. This is expressed as

$$A(t+\Delta t) - A(t) = N(t)\Delta t - [d(t, t+\Delta t) - d(t)]A_D + [b(t, t+\Delta t) - b(t)]A_B. \tag{2.28}$$

Dividing (2.28) by Δt and taking $\Delta t \to 0$, we get

$$\lim_{\Delta t \to 0} \frac{A(t+\Delta t) - A(t)}{\Delta t} = N(t) - \lim_{\Delta t \to 0}\left[\frac{d(t+\Delta t) - d(t)}{\Delta t}\right] A_D + \tag{2.29}$$

$$\lim_{\Delta t \to 0}\left[\frac{b(t+\Delta t) - b(t)}{\Delta t}\right] A_B$$

$$= N(t) - \frac{d}{dt}d(t)A_D + \frac{d}{dt}b(t)A_B. \tag{2.30}$$

In (2.30), $\frac{d}{dt}d(t)$ and $\frac{d}{dt}b(t)$ are changes due to death and birth numbers, A_D is the average death rate, and A_B is the average birth rate during $[t, t+\Delta t]$. We assume

$$\lim_{\Delta t \to 0}\left[\frac{d(t+\Delta t) - d(t)}{\Delta t}\right] \neq 0,$$

$$\lim_{\Delta t \to 0}\left[\frac{b(t+\Delta t) - b(t)}{\Delta t}\right] \neq 0.$$

Both birth and death rates play an important role in population aging analysis [19–23]. Let $A(x, t)$ indicate the age of the population who are at age x at time t, and then the partial differential equation that governs the dynamics of aging over all the ages $[0, \omega]$ would be

$$\frac{\partial A(x,t)}{\partial x} = \kappa \frac{\partial A(x,t)}{\partial t}, \tag{2.31}$$

where κ is defined as

$$\kappa = \int_x x \left(\frac{A(x,t)}{A(t)} \right) dx. \tag{2.32}$$

Even for stability analysis of a population model, the age-specific birth and death rates play an important role. For example, see [19].

Example 2.1 Suppose 2, 1, and 3 are the death durations in months among $N(t)$ and among newborns during $[t, t + \Delta t]$ combined;4 then the total time lost due to deaths by these three individuals is 3×2 months $= 6$ months.

Next, we provide an example of the average age gained due to births during the same interval.

Example 2.2 Suppose there are three births that occurred during the interval that completed 4, 1, and 4 months, respectively, by the end of $[t, t + \Delta t]$; then the total duration added by these births is $= 3 \times 3$ months $= 9$ months.

Suppose $N(t) = 10$ and $A(t) = 55$ years. If there are no births and deaths during next two years, then $A(t + 2) = 75$ years.

A population aging model with flexible birth rates will lead to a more realistic analysis of the aged population. The stability properties of the model need to be further understood. We leave Sect. 2.3 with the above preliminary revised aging model. For a general reading on the mathematics of aging models, refer to [18, 21, 24, 25].

2.4 Okubo Diffusion Model

In this section, we derive a PDE model of the form

$$\frac{\partial p(x,t)}{\partial t} = \alpha_1 \frac{\partial^2 p(x,t)}{\partial x^2} - \alpha_2 \frac{\partial p(x,t)}{\partial x}, \tag{2.33}$$

as proposed by ecologist Akira Okubo. Here, $p(x, t)$ represents the probability that an individual at a position x on the xt−plane (space-time plane) is located at that position at time t. The parameters α_1 and α_2 describe the diffusion model. Let θ

denote the probability that the individual moves to the right, while $(1-\theta)$ represents the probability that the individual moves to the left. If an individual always moves in one direction (either left or right), then $p(x, t+\Delta t)=0$ for a small change Δt in time. We develop the model within a random walk framework by expressing $p(x, t+\Delta t)$ in terms of $p(x-\Delta x, t)$ and $p(x+\Delta x, t)$. That is,

$$p(x, t+\Delta t)=\theta p(x-\Delta x, t)+(1-\theta) p(x+\Delta x, t). \tag{2.34}$$

This framework in (2.34) resembles the birth and death process we will explore in Chap. 6, where either one birth or one death occurs at each time step. In the birth and death model, the population can grow by one individual with each birth or decrease by one individual with each death during a short time interval. Let us rewrite the right side of (2.34) in a different way as follows:

$$\begin{aligned} p(x, t+\Delta t)= & \frac{1}{2} p(x-\Delta x, t)-\left(\frac{1}{2}-\theta\right) p(x-\Delta x, t)+\frac{1}{2} p(x+\Delta x, t)+ \\ & \left(\frac{1}{2}-\theta\right) p(x+\Delta x, t) . \end{aligned} \tag{2.35}$$

Adding a term $-p(x, t)$ on both sides of (2.35) and dividing by Δt, we get

$$\begin{aligned} \frac{p(x, t+\Delta t)-p(x, t)}{\Delta t} & =\frac{1}{2 \Delta t}\{p(x-\Delta x, t)+p(x+\Delta x, t)-2 p(x, t)\}+ \\ & \frac{\phi}{\Delta t}\{p(x+\Delta x, t)-p(x-\Delta x, t)\} \text { (where } \phi=(\frac{1}{2}-\theta)) \\ & =\frac{(\Delta x)^2}{2 \Delta t}\left\{\frac{p(x-\Delta x, t)+p(x+\Delta x, t)-2 p(x, t)}{(\Delta x)^2}\right\}- \\ & \frac{2 \phi \Delta x}{\Delta t}\left\{\frac{p(x+\Delta x, t)-p(x-\Delta x, t)}{-2 \Delta x}\right\} \\ & =\alpha_1\left\{\frac{p(x-\Delta x, t)+p(x+\Delta x, t)-p(x, t)}{(\Delta x)^2}\right\}- \\ & \alpha_2\left\{\frac{p(x+\Delta x, t)-p(x-\Delta x, t)}{2 \Delta x}\right\}, \end{aligned} \tag{2.36}$$

where $\alpha_1=\frac{(\Delta x)^2}{2 \Delta t}$, and $\alpha_2=-\frac{2 \phi \Delta x}{\Delta t}$. Taking $\Delta t \rightarrow 0$, $\Delta x \rightarrow 0$ in (2.36), we arrive at the diffusion model in (2.33). When $\alpha_2=0$, the quantity ϕ becomes 0. This implies $\theta=\frac{1}{2}$. The revised diffusion model under the equal chance of moving to the left or to the right becomes

$$\frac{\partial p(x, t)}{\partial t}=\alpha_1 \frac{\partial^2 p(x, t)}{\partial x^2}. \tag{2.37}$$

When $\theta = 1$ in (2.34), we arrive at

$$p(x, t + \Delta t) = p(x - \Delta x, t). \tag{2.38}$$

Again adding a term $-p(x, t)$ on both sides of (2.38) and dividing by Δt, we get

$$\begin{aligned}\frac{p(x, t + \Delta t) - p(x, t)}{\Delta t} &= \frac{p(x - \Delta x, t) - p(x, t)}{\Delta t} \\ &= \frac{-2\Delta x}{\Delta t}\left\{\frac{p(x - \Delta x, t) - p(x, t)}{-2\Delta x}\right\}.\end{aligned} \tag{2.39}$$

When $\theta = 1$, we have $\phi = -\frac{1}{2}$, which implies $\alpha_2 = \frac{\Delta x}{\Delta t}$. Substituting $\alpha_2 = \frac{\Delta x}{\Delta t}$ in (2.39), we get

$$\frac{p(x, t + \Delta t) - p(x, t)}{\Delta t} = -2\alpha_2\left\{\frac{p(x - \Delta x, t) - p(x, t)}{-2\Delta x}\right\}. \tag{2.40}$$

Taking $\Delta t \to 0$, in (2.40), the modified Okubo diffusion model under $\theta = \frac{1}{2}$ becomes

$$\frac{\partial p(x, t)}{\partial t} = -\alpha_2 \frac{\partial p(x, t)}{\partial x}. \tag{2.41}$$

Okubo's model has played an important role in the mathematical ecological dynamics. This work blends the ideas of random walks and population dynamics. For further understanding of the model, advances, and applications, refer to [26–29].

2.5 Exercises

Exercise 2.3 Let $u(a, t)$ be the proportion of the population at age a and time t. Then show that the diffusion equation of the type

$$\frac{\partial u}{\partial t} = k\frac{\partial^2 u}{\partial a^2},$$

can be used to represent entry of births into and removal of deaths in age class a (where k is a proportionality constant). Assume births are higher than deaths. ***Hint.*** *For age $t > s$*, use

$$\int_x^y \frac{\partial u(a, t)}{\partial t} da = k\frac{\partial u(y, t)}{\partial a} - k\frac{\partial u(x, t)}{\partial a}.$$

Exercise 2.4 Let $P(t)$ be the population at time t. List the solutions of $P(t)$ in the following two models: $\frac{dP(t)}{dt^2} + K^2P(t) = 0$, and $\frac{dP(t)}{dt^2} - K^2P(t) = 0$.

Exercise 2.5 Given an initial value problem of $\frac{dP(t)}{dt} = KP(t)$, $P(0) = 1$, find an expression for $P_n(t)$ as $n \to \infty$. [Hint: Verify Picard's uniqueness theorem. Here $(P_n(t))$ is a sequence of functions, where $P_{n+1}(t) = P(0) + \int_{t_0}^{t} f(s, P_n(s))\, ds$.]

Exercise 2.6 Given an initial value problem of $\frac{dP(t)}{dt} = K[P(t)]^{1/4}$, $P(1) = 0$. How many solutions does this initial value problem will have? Find them.

References

1. Rao, A.S.R.S.: Incubation periods under various anti-retroviral therapies in homogeneous mixing and age-structured dynamical models: a theoretical approach. Rocky Mountain J. Math. **45**(3), 973–1031 (2015)
2. McKendrick, A.G.: Applications of mathematics to medical problems. Proc. Edinb. Math. Soc. **13**, 98–130 (1926)
3. Kermack W.O., McKendrick A.G.: Contributions to the mathematical theory of epidemics. II. —The problem of endemicity. Proc. R. Soc. Lond. A **138**, 55–83 (1932). http://doi.org/10.1098/rspa.1932.0171
4. Keyfitz, B.L., Keyfitz, N.: The McKendrick partial differential equation and its uses in epidemiology and population study. Math. Comput. Modell. **26**(6), 1–9 (1997)
5. Bacaer, N.: A Short History of Mathematical Population Dynamics. Springer London, London (2011)
6. Hoppensteadt, F.: Mathematical Theories of Populations: Demographics, Genetics and Epidemics. Society for Industrial and Applied Mathematics, Philadelphia, PA (1975)
7. Halder, J., Tumuluri, S.K.: A higher order numerical scheme to a nonlinear McKendrick–von Foerster equation with singular mortality. Appl. Numer. Math. **202**, 21–41 (2024)
8. Akushevich, I., Yashkin, A., Kravchenko, J., Fang, F., Arbeev, K., Sloan, F., Yashin, A.I.: A forecasting model of disease prevalence based on the McKendrick–von Foerster equation. Math. Biosci. **311**, 31–38 (2019)
9. Hastings, A.: McKendrick Von Foerster Models for Patch Dynamics. In: Busenberg, S., Martelli, M. (eds.) Differential Equations Models in Biology, Epidemiology and Ecology. Lecture Notes in Biomathematics, vol 92. Springer, Berlin, Heidelberg (1991). https://doi.org/10.1007/978-3-642-45692-3_13
10. von Foerster, H.: Some remarks on changing populations. In The Kinetics of Cell Prolifeera-tion, pp. 382–407. A.G. McK (1959)
11. Webb, G.F.: Theory of Nonlinear Age-Dependent Population Dynamics. Marcel Dekker, New York (1985)
12. Inaba, H.: Age-Structured Population Dynamics in Demography and Epidemiology. Springer (2017)
13. Murray, J.D.: Mathematical Biology. An introduction I. Interdisciplinary Applied Mathematics, 3rd edn, vol. 17. Springer, New York (2002)
14. Kot, M.: Elements of Mathematical Ecology, x+453 pp. Cambridge University Press, Cambridge (2001)
15. Yosprakob, T., Shyntar, A., Iworima, D.G., Edelstein-Keshet, L.: Modeling the growth and size distribution of human pluripotent stem cell clusters in culture. Bull. Math. Biol. **86**(8), Paper No. 96, 32 pp. (2024)

16. Mishra, P., Ponosov, A., Wyller, J.: On the dynamics of predator-prey models with role reversal. (English summary) Phys. D **461**, Paper No. 134100, 20 pp. (2024)
17. Bartłomiejczyk, A., Leszczyński, H., Matusik, M.: Straightened characteristics of McKendrick–von Foerster equation. J. Differ. Equ. **340**, 592–615 (2022)
18. Preston, S.H., Himes, C., Eggers, M.: Demographic conditions responsible for population aging. Demography **26**, 691–704 (1989)
19. Rao, A.S.R.S.: Population stability and momentum. Notic. Am. Math. Soc. **61**(9), 1062–1065 (2014). (https://www.ams.org/notices/201409/rnoti-p1062.pdf)
20. Krantz, S.G.: Real analysis and foundations. Fifth edition [of MR1210958]. Textbooks in Mathematics, xv+484 pp. CRC Press, Boca Raton, FL (2022). ISBN:978-1-032-10272-6; 978-1-032-12026-3; 978-1-003-22268-2
21. Lee, R.D.: Rethinking the evolutionary theory of aging: transfers, not births, shape senescence in social species. Proc. Natl. Acad. Sci. **100**(16), 9637–9642 (2003)
22. Lee, R., Zhou, Y.: Does fertility or mortality drive contemporary population aging? The revisionist view revisited. Popul. Dev. Rev., 285–301 (2017)
23. Arokiasamy, P., Bloom, D., Lee, J., Feeney, K., Ozolins, M.: Longitudinal aging study in India: Vision, design, implementation, and preliminary findings. In: Aging in Asia: Findings from New and Emerging Data Initiatives, 5. National Academies Press (US) (2012)
24. Tuljapurkar, S.: The final inequality: variance in age at death. In: Demography and the Economy, pp. 209–221. University of Chicago Press (2010)
25. Cox, D.R., Oakes, D.: Analysis of survival data. Monographs on Statistics and Applied Probability. Chapman & Hall, London (1984)
26. Okubo, A., Levin, S.A.: Diffusion and Ecological Problems: Modern Perspectives, 2nd edn. Interdisciplinary Applied Mathematics, 14. Springer, New York (2001)
27. Kim, Y.-J.; Kwon, O.; Li, F.: Global asymptotic stability and the ideal free distribution in a starvation driven diffusion. J. Math. Biol. **68**(6), 1341–1370 (2014)
28. Numfor, E., Hilker, F.M., Lenhart, S.: Optimal culling and biocontrol in a predator–prey model. Bull. Math. Biol. **79**, 88–116 (2017)
29. Kaur, H., Agnihotri, K., Melese, D.: The spatio-temporal study of a planktonic system having toxin producing and infected phytoplankton species. Differ. Equ. Dyn. Syst. **33**(1), 141–175 (2025)

Chapter 3
Life Table

The life table sometimes called the mortality table is one of the oldest mathematical models in demography. In demography and actuarial sciences, life tables are essential tools for data analysis and predictions. The life table is a stationary population model where the total births and deaths are identical, and the total population of a life table is constant. The life table is perhaps the oldest tool to compute the expectation of life of a cohort of births. It is unclear if understanding human behavior, society, and culture were part of the human evolution. It is still unclear if there was a method for estimating longevity of humans prior to the life table techniques. In this chapter, we will learn the methodologies of the computation of life table and the assumptions involved in constructing various columns of the life table. We use the standard notations to represent various columns of a life table.

In the previous chapter, we studied partial differential equation (PDE) models in population dynamics. These models, along with the ordinary differential equation (ODE) models introduced in Chap. 4, are examples of deterministic mathematical models. In deterministic modeling, the outcomes are entirely governed by the initial conditions and parameters, with no element of randomness or uncertainty.

The life table is another important example of a deterministic model in demography. Once the age-specific mortality rates are obtained from actual population data, all other columns in the life table, such as the number of survivors, deaths, person-years lived, and life expectancy, are constructed using fixed mathematical formulas. There are no uncertainty factors involved in this process.

In this chapter, we will explore the structure and mathematical foundation of life tables and examine how they serve as a reliable tool for analyzing mortality and survival patterns in populations.

A. S. R. Srinivasa Rao, *Mathematical Demography: Theory and Modeling*,
Mathematical Marvels: Texts and Monographs in the Spirit of CR Rao,
https://doi.org/10.1007/978-981-95-6129-2_3

3.1 Force of Mortality

Before one can start constructing various columns of a life table, it is essential to understand the *central death rates* m_a at each age a in the population in a year t for which a life table is constructed. The quantity m_a is also called the *age-specific mortality rate*. Computation of m_a is the first step, and the data to do this can be obtained from a population survey conducted in a year. These mortality rates are occasionally obtained from reliable secondary sources, for example, the census, etc. The formula for m_a is represented by

$$m_a = \frac{D_a}{P_a} \approx \frac{d_a}{L_a}, \tag{3.1}$$

where D_a is the number of deaths at age a (deaths among who reached age a and before reaching the age $a+1$), and P_a is the midyear population[1] at age a obtained for a year for which a life table is to be constructed. The midyear population is computed as an average value of the populations recorded at the beginning and at the end of the year. The quantities d_a and L_a are corresponding deaths and populations at age a of the life table. The assumption that $\frac{D_a}{P_a} \approx \frac{d_a}{L_a}$ is the first major assumption of bringing real-world data closer to the synthetic mortality pattern used in a life table. Soon, we will learn about constructing these two quantities. The force of mortality at age $a+t$ is defined as

$$\mu_{a+t} = \lim_{\Delta t \to 0} \frac{l_{a+t} - l_{a+t+\Delta t}}{\Delta t l_a}, \tag{3.2}$$

where l_{a+t} is the number surviving at age $a+t$ and is usually approximated by $l_{a+t} \approx l_a - td_a$ for $0 \leq t \leq 1$. For $t = 1$, we have $l_{a+1} = l_a - d_a$. Similarly, $l_{a+t+\Delta t}$ is the number surviving at age $a+t+\Delta t$. Using the force of mortality at age a denoted by μ_a and the survivors at age a denoted by l_a, the quantity m_a is expressed as follows:

$$m_a = \frac{\int_0^1 l_{a+t}\mu_{a+t}dt}{\int_0^1 l_{a+t}dt}. \tag{3.3}$$

[1] The midyear population at age a (i.e., $P_a^{(t)}$) for a year t is often computed through

$$P_a^{(t)} = \frac{P_a^{(t,\ \text{January 1})} + P_a^{(t,\ \text{December 31})}}{2},$$

where $P_a^{(t,\ \text{January 1})}$ and $P_a^{(t,\ \text{December 31})}$ are census or even estimated populations at age a for January 1 and December 31, respectively, for the year t.

Using (3.2), we can write μ_a as

$$\mu_a = \lim_{\Delta t \to 0} \frac{l_a - l_{a+\Delta t}}{\Delta t l_a} \approx -\frac{1}{l_a}\frac{dl_a}{dt}. \tag{3.4}$$

Numerical methods of approximation on functions[2] were used in the above approximation of μ_a. Using (3.4), we can write

$$\frac{dl_{a+t}}{dt} = -l_{a+t}\mu_{a+t}, \tag{3.5}$$

and

$$\int_0^1 l_{a+t}\mu_{a+t}dt = -\int_0^1 \frac{dl_{a+t}}{dt}dt = -\ l_{a+t}|_0^1 = l_a - l_{a+1} = d_a. \tag{3.6}$$

For practical purposes, once m_a is computed from the real data, in the second step, the values of probabilities of dying at age a (in the corresponding life table) denoted by q_a are computed using the formula

$$q_a = \frac{2m_a}{2 + m_a}. \tag{3.7}$$

The quantity q_a is interpreted as the probability of dying of an individual during the age $[a, a+1)$, i.e., those dying after attaining age a and before reaching the age $a+1$. The difference between m_a and μ_a is that m_a is the age-specific mortality rate (also sometimes referred to as age-specific mortality rate) of individuals in the age interval $[a, a+1)$, whereas μ_a is the instantaneous death rate exactly at age a. The survival probabilities at each age denoted by p_a are computed using $p_a = 1 - q_a$ for all a. Here p_a is interpreted as the probability that an individual of age a will survive for one more year, and it is expressed as $p_a = l_{a+1}/l_a$. This implies d_a in (3.1) can be computed using $q_a l_a$, that is, $d_a = q_a l_a = (1 - p_a)l_a = \left(1 - \frac{l_{a+1}}{l_a}\right)l_a$. The quantity L_a in (3.1) is called the life table population at age a bringing an important new concept called the *person-years lived* by individuals of age a during $[a, a+1)$. In other words, L_a is the total years lived by l_a individuals during $[a, a+1)$. The quantity L_a is defined as

$$L_a = \int_a^{a+1} l_s ds = \int_0^1 l_{a+t}dt. \tag{3.8}$$

[2] The derivative of a function $f(x)$ at a point $x = a$ is defined as $\frac{df(x)}{dx} = \lim_{\Delta t \to 0} \frac{f(a+\Delta t) - f(a)}{\Delta t}$ is approximated using numerical methods as $\frac{df(x)}{dx} \approx \frac{f(x+\Delta t) - f(x-\Delta t)}{2\Delta t}$.

This implies

$$\frac{dL_a}{da} = l_{a+1} - l_a = -d_a \text{ (from (3.6))}. \quad (3.9)$$

The notation d_a is not to be confused with da. Using the trapezoidal rule of numerical integration[3] to the formula in (3.8), we write L_a as

$$L_a = \frac{l_a + l_{a+1}}{2} = \frac{l_a + l_a + l_{a+1} - l_a}{2} = l_a - \frac{1}{2} d_a.$$

Assuming uniform distribution of deaths during $[a, a+1)$, the quantity L_a can be approximated by $L_a \approx l_{a+\frac{1}{2}}$, where $l_{a+\frac{1}{2}}$ is the number of survivors at the age $a+\frac{1}{2}$, that is, survivors in the midyear during $[a, a+1)$. The total person-years that will be lived by individuals of age a until their death is denoted by T_a and is defined by

$$T_a = \int_a^\infty l_s ds = \int_0^\infty l_{a+t} dt \approx \sum_{s=a}^{\omega} L_s. \quad (3.10)$$

This implies $T_0 = \sum_{s=0}^{\omega} L_s$, $T_1 = \sum_{s=1}^{\omega} L_s = T_0 - L_0$, and so on.

3.2 The Survival Pattern and the Life Expectancy

Once the death numbers d_a at each age are constructed using the relations explained in the previous section, the aim of a life table is to construct l_a values for all the single ages from 0 to a maximum age of survival which is usually denoted by ω. In a way, the l_a column is crucial to understand the survival patterns of a given population starting with a synthetic cohort number $l_0 = 100{,}000$. The l_0 is also called the *radix* of a life table. The value of l_1 is obtained from the relation $l_1 = l_0 - d_0$ and the value of l_2 is obtained from $l_2 = l_1 - d_1$, and so on. The values of d_a are obtained from what we derived above as follows: $d_0 = q_0 l_0$, $d_1 = q_1 l_1$, and so on The quantity $l_\omega = 0$. The function l_a is a nonincreasing function of a and it can be represented by a graph shown in Fig. 3.1.

[3] Let $[a, b]$ be a real-valued interval in which an integral of a function $f(x)$ is defined. Let $[a, b]$ be partitioned into k intervals such that $[a, b] = [a = a_0, a_1) \bigcup_{i=2}^{k-1} (a_{i-1}, a_i) \cup (a_{k-1}, a_k = b]$, where $a_1 - a_0 = a_i - a_{i-1} = b - a_k = \Delta a_i$ for $i = 2, 3, \ldots, k-1$. The trapezoidal rule says,

$$\int_a^b f(x)dx \approx \left[\frac{f(a) + f(a_1)}{2} + \frac{1}{2} \sum_{i=2}^{k-1} \{f(a_{i-1}) + f(a_i)\} + \frac{f(a_{k-1}) + f(a_k)}{2} \right] \Delta a_i.$$

Since we are constructing a single-age life table, the interval $[a, a+1)$ is not divided further, and our $\Delta a_i = 1$. By the trapezoidal rule formula, $L_a = \int_a^{a+1} l_s ds \approx \frac{l_a + l_{a+1}}{2}.(a+1-a) = \frac{l_a + l_{a+1}}{2}$.

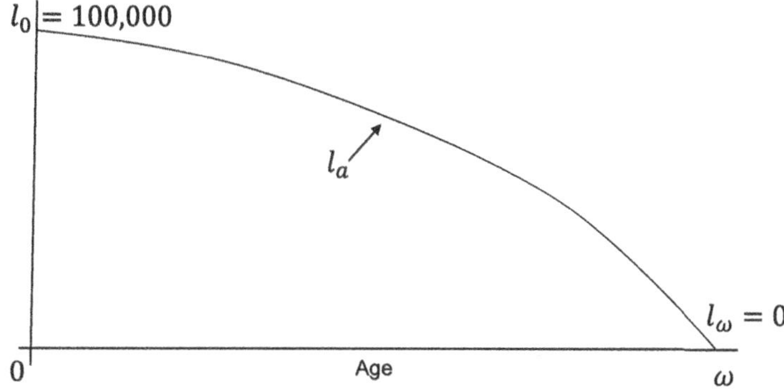

Fig. 3.1 A sample l_a curve. The l_a column in a life table is a nonincreasing function, i.e., $l_a \geq l_{a+1}$ for $a = 0, 1, \ldots, \omega$. The initial value of a cohort size is $l_0 = 100{,}000$ or any suitable size of newborn individuals at time t

Using the l_a column, we construct L_a, and using L_a, we construct the column T_a. The column T_a combined with the column l_a will be used in the computation of the last column of the life table, called the life expectancy column, denoted by e_a^0, i.e., expectation of life at age a or an average future time to be lived by an individual of age a of a life table population and is represented by

$$e_a^0 = \frac{1}{l_a} \int_0^\infty l_{a+t} dt. \tag{3.11}$$

The first value of the column e_a^0 for $a = 0$ is denoted by e_0^0. The quantity e_0^0 is known as the life expectancy at birth. Earlier in Chap. 1, we saw the importance of e_0^0 in understanding the birth rate in a stable population. For the sake of understanding how we practically construct a single-age life table based on the formulae derived above, a hypothetical single-age life table is computed in this section for a population (say, human, animal, insect, or bird population cohort of size $l_0 = 25$). See Table 3.1. A single-age life table is also sometimes called a single-year life table. Later, we will learn about abridged life tables (non-single-age life tables). See Sect. 3.4. As an example, let us construct a single-age life table for a population that has experienced age-specific mortality rates as follows: $m_0 = 0.174, m_1 = 0.1$, $m_2 = 0.171, m_3 = 0.207$, $m_4 = 0.167, m_5 = 0.095$, $m_6 = 0.353, m_7 = 0.545$, $m_8 = 0.6$, $m_9 = 0.667, m_{10} = 0.667$, $m_{10} = 0.690, m_{11} = 0$. We previously noted that the A is based on real population data, and these values serve as the main input for the life table through column B in the corresponding life table.

From these m_a values, we compute q_a of Table 3.1 using the formula in (3.7). Next, d_a values are computed using $d_a = l_a q_a$ described earlier. For example, $d_0 = l_0 q_0 = 25 \times 0.16 = 4$, and so on. The quantity L_0 is computed in Table 3.1 by $L_0 = l_0 - \frac{1}{2} d_0 = 25 - \frac{1}{2} \times 4 = 23$, $L_0 = l_1 - \frac{1}{2} d_1 = 21 - \frac{1}{2} \times 2 = 20$ and so on.

The last values of the columns L_a and T_a for $a = 11$ in Table 3.1 are adjusted using standard assumptions like $L_{11} = d_{11}/m_{11}$. This adjustment was done to

Table 3.1 A hypothetical life table for a cohort size of 25 animal, insect, or bird population. See the text in Sect. 3.2 for the assumptions related to the last row of the life table

Age	l_a	d_a	q_a	L_a	T_a	e_a^0
0	25	4	0.160	23	117	4.7
1	21	2	0.095	20	94	4.5
2	19	3	0.158	17.5	74	3.9
3	16	3	0.188	14.5	56.5	3.5
4	13	2	0.154	12	42	3.2
5	11	1	0.091	10.5	30	2.7
6	10	3	0.300	8.5	19.5	2.0
7	7	3	0.429	5.5	11	1.6
8	4	2	0.462	3.1	5.5	1.4
9	2	1	0.500	1.6	2.4	1.1
10	1	1	0.513	0.5	0.8	0.7
11	0	0	0.0	0	0	0
		$\sum_{a=0}^{11} d_a = 25$		$\sum_{a=0}^{11} L_a = 117$		

make sure $L_{11} \approx T_{11}$. This additional last row assumption was not part of the mathematical formulae derived above but was introduced often to make $L_\omega \approx T_\omega$ and to nicely complete the life table. Further on assumptions on the last row in the life table can be seen in [1–6]. Using (3.11), one can construct life expectancy at birth of a cohort or life expectancy at any age. See the paragraph at the end of this chapter where further references are provided on actual life tables for various countries and subpopulations in the world.

3.3 Relationship Between m_a and μ_a

It becomes useful sometimes to obtain μ_a values from estimated m_a. We have seen a relation between m_a, d_a, and L_a in (3.1) and a relation between the rate of change in L_a and d_a in (3.9). Using these two sets of relations, we can write

$$m_a = \frac{d_a}{L_a} = -\frac{1}{L_a}\frac{dL_a}{da}. \tag{3.12}$$

Assuming uniform distribution of deaths, we saw earlier that $L_a \approx l_{a+\frac{1}{2}}$. This approximation will help to write m_a in (3.12) as

$$m_a = -\frac{1}{l_{a+\frac{1}{2}}}\frac{dl_{a+\frac{1}{2}}}{da} = \mu_{a+\frac{1}{2}}. \tag{3.13}$$

Remark 3.1 (Gompertz's Law and Makeham's Law) Gompertz's law of mortality published in 1825 [7] assumed μ_a to follow a geometric progression. According to Gompertz's law $\mu_a = Bc^a$, where B and c are positive constants. Makeham's law of mortality is expressed as $\mu_a = A + Bc^a$ for constants A, B, and c [8]. As mentioned previously, the life table aims to obtain a pattern for l_a such that one can estimate life expectancy at each age a using the column T_a. Using Gompertz's and Makeham's laws, one can obtain a functional form of l_a described below:

When $\mu_a = Bc^a$,

$$\begin{aligned}\int_0^a \mu_s ds &= \int_0^a Bc^s ds = B\int_0^a c^s ds = B\left.\frac{c^s}{\log c}\right|_0^a \\ &= B\left(\frac{c^a}{\log c} - \frac{1}{\log c}\right) \\ &= \frac{B\left(c^a - 1\right)}{\log c}. \end{aligned} \tag{3.14}$$

Using (3.4), we can write

$$\begin{aligned}\int_0^a \mu_s ds &= \int_0^a \left(-\frac{1}{l_s}\frac{dl_s}{ds}\right) ds \left(= \int_0^a \left(-\frac{d\log l_s}{ds}\right) ds = -\log l_s|_0^a\right) \\ &= -\log\frac{l_a}{l_0}. \end{aligned} \tag{3.15}$$

This implies

$$l_a = l_0 \exp\left(-\int_0^a \mu_s ds\right). \tag{3.16}$$

From (3.14) and (3.16), we can express l_a using Gompertz's law as

$$l_a = l_0 \exp\left\{-\frac{B\left(c^a - 1\right)}{\log c}\right\}. \tag{3.17}$$

When $\mu_a = A + Bc^a$, we have

$$\begin{aligned}\int_0^a \mu_s ds &= \int_0^a (A + Bc^s) ds = \int_0^a A ds + B\int_0^a c^s ds = Aa + B\left.\frac{c^s}{\log c}\right|_0^a \\ &= Aa + B\left(\frac{c^a}{\log c} - \frac{1}{\log c}\right) \\ &= Aa + \frac{B\left(c^a - 1\right)}{\log c}. \end{aligned} \tag{3.18}$$

From (3.18) and (3.16), we can express l_a using Makeham's law as

$$l_a = l_0 \exp\left[-\left\{Aa + \frac{B\left(c^a - 1\right)}{\log c}\right\}\right]. \tag{3.19}$$

The l_a pattern in (3.17) is the pattern of survival of individuals based on Gompertz's law, and the l_a pattern in (3.19) is the pattern of survival of individuals based on Makeham's law.

3.4 Abridged Life Table Functions

Previously, we have introduced life table construction for single-year age intervals like the function p_a indicating the probability of survival of an individual who is at age a to survive for one more year, that is, $p_a = l_{a+1}/l_a$. We saw the construction of life table in Table 3.1 for all the ages of a population. Now, let us define life table functions that provide information on more than 1-year age interval, for example, ${}_np_a = l_{a+n}/l_a$ for $n = 1, 2, 3, \ldots, \omega$. The value of the quantity ${}_np_a$ indicates the probability of survival of an individual of age a to survive for n more years from age a. Similarly, the function ${}_nq_a$ is expressed as

$$\begin{aligned}
{}_nq_a = 1 - {}_n p_a &= \frac{l_a - l_{a+n}}{l_a} \\
&= \frac{-\int_a^{a+n} \frac{dl_s}{ds} ds}{l_a} \left(= \frac{1}{l_a} \int_0^n l_{a+s}\mu_{a+s} ds\right) \quad (3.20) \\
&= \frac{1}{l_a} \int_a^{a+n} l_s \mu_s ds. \quad (3.21)
\end{aligned}$$

In (3.20), we used relations (3.5) and (3.6). The number of deaths within the interval $[a, a+n)$ is $l_a - l_{a+n} = {}_nd_a$. Using (3.21), we can write ${}_nq_a = \int_0^n {}_sp_a \mu_{a+s} ds$. The number of survivors at age $a + n$ is expressed using (3.16) as

$$l_{a+n} = l_0 \exp\left(-\int_0^{a+n} \mu_s ds\right). \tag{3.22}$$

This gives us

$${}_np_a = \frac{l_{a+n}}{l_a} = \exp\left(-\int_a^{a+n} \mu_s ds\right).$$

For practical purposes, ${}_nq_a$ is computed from the age-specific mortality rates within the age $[a, a+n)$ and linear death assumption within $[a, a+n)$ using the relation

$$ {}_nq_a = \frac{2n.{}_nm_a}{2 + n.{}_nm_a}. \tag{3.23} $$

The person-years lived by l_a during $[a, a+n)$ with the linearity assumption on deaths denoted by ${}_nL_a$ is obtained using (3.8)

$$ {}_nL_a = \int_a^{a+n} l_s ds = \int_0^n l_{a+t} dt, \tag{3.24} $$

and it is approximated using the trapezoidal rule described earlier by ${}_nL_a \approx \frac{n}{2}(l_a + l_{a+n})$. These abridged quantities of ${}_nL_a$ will be utilized in population projections. Instead of a linear assumption on l_a, one can construct a life table with nonlinear curves to model l_a and obtain other columns using the relationships previously built in this chapter (Sects. 3.2 and 3.3). We saw two examples that used Gompertz and Makeham laws to model l_a. Let us now see another example to model μ_a by considering μ_a as a polynomial (which was not considered earlier by others). Let us obtain an expression for ${}_np_a$ using the new polynomial form for μ_a in Example 3.2. To obtain ${}_np_a$, we need to express l_{a+n} and l_a for a given form of μ_a in Example 3.2. Let Δt be a small time interval in which survival of an individual of age a is measured; then the force of mortality μ_a is also defined using the function ${}_{\Delta t}q_a$—probability of dying as

$$ \mu_a = \lim_{\Delta t \to 0^+} \frac{{}_{\Delta t}q_a}{\Delta t}. \tag{3.25} $$

Later in Chap. 8, we will see that the force of mortality of the life table plays an important role in understanding central quantities in the subject survival analysis within the field of statistics. The force of mortality is treated there as a hazard rate.

Example 3.2 Let us consider a 4th degree polynomial[4] to model the force of mortality at age s given by

$$ \mu_s = b_4 s^4 + b_3 s^3 + b_2 s^2 + b_1 s + b_0, \tag{3.26} $$

where b_0, b_1, b_2, b_3, and b_4 are real numbers. Let us construct l_{a+n} for μ_s in (3.26) using (3.22) as follows:

$$ \begin{aligned} \int_0^{a+n} \mu_s ds &= \int_0^{a+n} \left(b_4 s^4 + b_3 s^3 + b_2 s^2 + b_1 s + b_0 \right) ds \\ &= b_4 \int_0^{a+n} s^4 ds + b_3 \int_0^{a+n} s^3 ds + b_2 \int_0^{a+n} s^2 ds + \end{aligned} $$

[4] Let n be some integer greater than or equal to 0. A polynomial $p(t)$ with coefficients a_0, a_1, $a_2, \ldots, a_n$ on the real line and with a variable t is represented by $p(t) = a_n t^n + a_{n-1} t^{n-1} + \ldots + a_1 t + a_0$.

$$b_1 \int_0^{a+n} s ds + b_0 \int_0^{a+n} ds$$

$$= b_4 \left.\frac{s^5}{5}\right|_0^{a+n} + b_3 \left.\frac{s^4}{4}\right|_0^{a+n} + b_2 \left.\frac{s^3}{3}\right|_0^{a+n} + b_1 \left.\frac{s^2}{2}\right|_0^{a+n} +$$

$$b_0 \; s|_0^{a+n} + C,$$

(where C is the constant of integration)

$$= C + \sum_{i=0}^{4} b_i \frac{(a+n)^{i+1}}{i+1}. \tag{3.27}$$

Therefore, substituting the expression for $\int_0^{a+n} \mu_s ds$ obtained in (3.27) in l_{a+n} of (3.22), we get

$$l_{a+n} = l_0 \exp\left[-C - \sum_{i=0}^{4} b_i \frac{(a+n)^{i+1}}{i+1}\right]$$

$$= l_0 D_1 \exp[-\sum_{i=0}^{4} b_i \frac{(a+n)^{i+1}}{i+1}], \tag{3.28}$$

where $D_1 = e^{-C}$. Similarly, we construct l_a for (3.26) as follows:

$$\int_0^a \mu_s ds = \int_0^a \left(b_4 s^4 + b_3 s^3 + b_2 s^2 + b_1 s + b_0\right) ds$$

$$= K + \sum_{i=0}^{4} b_i \frac{a^{i+1}}{i+1} \tag{3.29}$$

(where K is the constant of integration)

Substituting (3.29) in (3.16), we get

$$l_a = l_0 \exp\left[-K - \sum_{i=0}^{4} b_i \frac{a^{i+1}}{i+1}\right]$$

$$= l_0 D_2 \exp\left[-\sum_{i=0}^{4} b_i \frac{a^{i+1}}{i+1}\right], \tag{3.30}$$

where $D_2 = e^{-K}$. The quantity ${}_n p_a$ can be expressed using (3.28) and (3.30) as follows:

$$
\begin{aligned}
{}_n p_a &= \frac{D_1 \exp[-\sum_{i=0}^{4} b_i \frac{(a+n)^{i+1}}{i+1}]}{D_2 \exp\left[-\sum_{i=0}^{4} b_i \frac{a^{i+1}}{i+1}\right]} \\
&= \frac{D_1}{D_2} \exp\left[-\left\{\sum_{i=0}^{4} \frac{b_i}{i+1}(a+n)^{i+1} - a^{i+1}\right\}\right]
\end{aligned}
$$

$$
\begin{aligned}
&= \frac{D_1}{D_2} \exp\left[b_0 n + \frac{b_1}{2}\left(2an + n^2\right) + \frac{b_2}{3}\left(3a^2 n + 3an^2 + n^3\right)\right| \\
&+ \frac{b_3}{4}\left(4a^3 n + 6a^2 n^2 + 4an^3 + n^4\right) \\
&+ \frac{b_4}{5}\left(5a^4 n + 10a^3 n^2 + 10a^2 n^3 + 5an^4 + n^5\right)\Bigg]. \qquad (3.31)
\end{aligned}
$$

Example 3.3 Let $\mu_s = Bs^p$ for a real number B and a positive integer p. Let us now find an expression for ${}_n p_a$.

$$
\begin{aligned}
\int_a^{a+n} \mu_s ds &= B \int_a^{a+n} s^p ds \\
&= \frac{B}{p+1}\left\{(a+n)^{p+1} - a^{p+1}\right\} + C,
\end{aligned}
$$

where C is the constant of integration. Now, the expression for ${}_n p_a$ will be

$$
\begin{aligned}
{}_n p_a &= \exp\left[-\frac{B}{p+1}\left\{(a+n)^{p+1} - a^{p+1}\right\} - C\right] \\
&= D \exp\left[-\frac{B}{p+1}\left\{(a+n)^{p+1} - a^{p+1}\right\}\right], \qquad (3.32)
\end{aligned}
$$

where $D = e^{-C}$.

The above two model examples of μ_s obtaining corresponding expressions for ${}_n p_a$ are helpful in several contexts of life table constructions.

3.5 Population Projections

Once the life table constructions and formulae for computations of person-years both for single-ages and abridged ages are understood, they can be applied in population projections. First, we show how to compute the cohort survival probability for a cohort of a given age group from life table data and then show how to use it to project census or actual population.

We saw previously that ${}_nL_a$ in a life table represents the person-years lived during age interval $[a, a+n)$ by individuals who attained age a or the total number of years lived by individuals of age a during $[a, a+n)$ for $a = 0, 1, \ldots, \omega$. The quantity ${}_5L_a$ is the total number of years lived by individuals who attained age a during the next five years. The quantity ${}_nL_0$ is the total number of years lived by the newborn individuals during $[0, n)$ for $n = 1, 2, ...$ The quantity

$$\frac{{}_5L_{a+5}}{{}_5L_a} \text{ for } a = 0, 5, ... \tag{3.33}$$

represents the survival probability of ${}_5L_a$ cohort of population to survive for five more years. Let these life table quantities be computed for a year t using female population data. Let ${}_5P_a^{(t)}$ be the female census population for the age group $[a, a+5)$ in the year t. Then, the quantity

$${}_5P_a^{(t)} \frac{{}_5L_{a+5}^{(t)}}{{}_5L_a}$$

represents the number of individuals of ${}_5P_a^{(t)}$ who survived until the year $t+5$. The survivors of ${}_5P_a^{(t)}$ will be in the age $[a+5, 5)$, which is represented by ${}_5P_{a+5}^{(t+5)}$. The quantity ${}_5P_{a+5}^{(t+5)}$ is the projected population and is represented by

$${}_5P_{a+5}^{(t+5)} = {}_5P_a^{(t)} \frac{{}_5L_{a+5}^{(t)}}{{}_5L_a}. \tag{3.34}$$

In the projected population expressed in (3.34), the projected births during $[t, t+5)$ are missing. The projected births during $[t, t+5)$ are computed through

$$\frac{5}{2}\left[{}_5P_a^{(t)} + {}_5P_{a-5}^{(t)} \frac{{}_5L_a^{(t)}}{{}_5L_{a-5}}\right] \frac{{}_5B_a^{(t)}}{{}_5P_a^{(t)}}, \tag{3.35}$$

where ${}_5B_a^{(t)}$ is the number of births in the year t born to the mothers of age $[a, a+5)$. We need to sum this quantity in (3.35) over all the ages for $a \in [\alpha, \beta]$, where α is the lower reproductive age and β is the upper reproductive age to get the total projected births by all the women in the reproductive ages. The total projected births needed to be multiplied by the quantity ${}_5L_0^{(t)}$ to obtain the net total births during $[t, t+5)$, say ${}_5P_0^{(t)}$. The net population in the age group $[0, 5)$ in the year t is represented as

$${}_5P_0^{(t)} = \frac{5}{2}\left[{}_5P_a^{(t)} + {}_5P_{a-5}^{(t)} \frac{{}_5L_a^{(t)}}{{}_5L_{a-5}}\right] \frac{{}_5B_a^{(t)}}{{}_5P_a^{(t)}} {}_5L_0^{(t)}.$$

We repeat the procedure for all the age groups and for both the male and female populations because the life tables are separately constructed for both populations.

3.6 Life Table Ideas for Social Sciences

In addition to constructing life tables to understand survival rates and the impact of deaths at various ages, it is also possible to create life tables to analyze other types of decrement data. For instance, we can develop the column l_a based on a cohort of children (l_0) who are starting school, using dropout data at each age. These tables may vary by region or country. By utilizing such school decrement tables, one can estimate average school years and remaining years of education, among other metrics.

Furthermore, other social studies data, such as marriage rates and employment statistics, can be analyzed using the principles of life tables. Life tables can also be developed based on morbidity data from various health conditions to estimate average healthy life years remaining. These decrement tables, which are designed to study multiple independent decrements, are referred to as multiple decrement tables and are introduced in Chap. 8.

Several textbooks and reference books on the construction of life tables are available [1–3, 9–11]. Uniform assumption of deaths in an interval is a frequently used method in computation. There are other types of assumptions and associated life tables constructed, for example, King's method, Reed–Merrell method, Chiang's method, etc. In Example 3.2, we saw a new form of μ_a by considering it as a 4th degree polynomial.

Life tables play an important role in the insurance industry and life contingencies. There are several country-level life tables published periodically [4–6, 12–16]. For further reading materials on life tables and formal relationships between various columns of a life table, refer to [17–20]. For additional references on Gompertz law and Gompertz–Makeham-type distributions, refer to [21–23]. Instead of total population life tables by male and female, one can also construct life tables for selected subpopulations. The life table analysis is an invaluable tool in mortality and morbidity analysis and insurance statistics of humans, insurance statistics analysis of the automobile industry, etc.; for example, see [24–32]. More on life table functions, we will study in the chapter on multiple decrement tables.

3.7 Exercises

Exercise 3.4 Let $\int_0^n l_{a+t}dt$ be the person-years lived during the age interval $[a, a+n]$. Derive an expression for the life expectancy at birth by assuming uniform death distribution. Here $l_{a+n} = l_0 \exp\left(-\int_0^{a+n} \mu_s ds\right)$, where l_0 is the radix.

Exercise 3.5 Construct all the major columns of a life table using the data in Table 3.2. Assume uniform death distribution.

Exercise 3.6 Obtain a relationship between the following quantities measuring at age a taking two quantities at a time: the force of mortality μ_a, the number of deaths

Table 3.2 l_a and d_a columns of a life table

Age	l_a	d_a	Age	l_a	d_a
0	41	10	9	7	1
1	31	6	10	6	2
2	25	5	11	4	2
3	20	3	12	2	1
4	17	3	13	1	1
5	14	2	14	0	0
6	12	1			$\sum_{a=0}^{14} d_a = 41$
7	11	2			
8	9	2			

d_a, and age-specific mortality rate m_a. In obtaining a relationship between the two, the third quantity can be considered as an intermediate relationship.

Exercise 3.7 Given that there is a constant force of mortality of 0.05 for all ages in a population, find (i) an individual aged 3 would die between 4 and 5, (ii) expectation of life at birth, and (iii) probability of survival at age 51.

Exercise 3.8 Given $\mu_a = \frac{1}{n^2} a^n \; n = 1, 2, ...,$ obtain an expression for ${}_n p_a$.

Exercise 3.9 (i) Show that $\frac{d}{dt}{}_t p_a \; =_t \; p_a \left(\mu_a - \mu_{a+t}\right)$. (ii) Show that $e_a = p_a \left(1 + e_{a+1}\right)$ by assuming $e_a = \frac{l_{a+1} + ... l_\omega}{l_1}$.

Exercise 3.10 Show that $\int_{-1}^{1} \mu_{a+t} dt = -\log p_{a-1}(1) - \log p_a(1)$.

References

1. Keyfitz, N., Caswell, H.: Applied Mathematical Demography, 3rd edn, xxi+441 pp. Springer Texts in Statistics. Springer, New York (2013)
2. Carey, J.R., Roach, D.: Biodemography. An Introduction to Concepts and Methods. Princeton University Press, Princeton (2019)
3. Ramkumar, R.: Technical Demography. New Age International Press, New Delhi (1986)
4. SRS: Sample Registration System - Abridged Life Tables 2016–2020 (2022). https://censusindia.gov.in/census.website/data/SRSALT
5. SRS: Sample Registration System - Abridged Life Tables 2014–2018 (2020). https://censusindia.gov.in/census.website/data/SRSALT
6. U.S. State Life Tables, 2020: Arias, Elizabeth; Xu, Jiaquan; Tejada-Vera, Betzaida; Murphy, Sherry L.; Bastian, Brigham, Series: NCHS National Vital Statistics Reports (2022). https://stacks.cdc.gov/view/cdc/118271
7. Gompertz , B.: On the nature of the function expressive of the Law of Human Mortality and on a new mode of determining the value of life contingencies. Phil. Trans. Royal Soc. **36**, 513–585 (1825)
8. Makeham, W.M.: On the law of mortality. J. Inst. Actuar. **13**(6), 325–358 (1867)
9. Misra, B.D.: An Introduction to the Study of Population. South Asian Publishers, New Delhi (1982)

10. Shryock, H., Siegel, J., Stockwell, E.: The Methods and Materials of Demography (Condensed Edition). Academic Press, New York (1976)
11. Keyfitz, N.: Introduction to the Mathematics of Population. Addison-Wesley, Reading, MA (1968)
12. Lahiri, S.: Survival probabilities from 5-year cumulative life table survival ratios (Tx + 5/Tx): some innovative methodological investigations. In: Rao, C.R. (ed.) Handbook of Statistics: Integrated Population Biology and Modelling, pp 481–542. Elsevier, Amsterdam (2018)
13. Ponnapalli, K.M.: A re-look at the methods for decomposing the difference between two life expectancies at birth. Sankhyā: Indian J. Stat. Ser. B, 283–309 (2008)
14. Abridged Life Tables for Japan: Life tables for Japan. Ministry of Health, Labour and Welfare (2022). https://www.mhlw.go.jp/
15. García-de-Alba-Verduzco, J.E., García-de-Alba-Verduzco, J.E., López-Elizalde, R., García-de-Alba-García, J.E.: Table of life for beneficiaries of the Institute of Security and Social Services of State Workers (ISSSTE), Mexico 2021. Cir. Cir. 2024 **92**(5), 594–602 (2021). English. https://doi.org/10.24875/CIRU.23000321. PMID: 39401776
16. Aries, E.: United States Life Tables, 2008. Natl. Vital Stat. Rep. **61**, 3 (2012)
17. Krishnamoorthy, S., Potter, R.G., Pickard, D.K. Population momentum: its relation to the moments of replacement level fertility. Math. Biosci. **53**(1–2), 41–51 (1981)
18. Jordan C.W.: Life Contingencies, 2nd edn. The Society of Actuaries, Chicago (1967)
19. Wachter, K.W.: Essential Demographic Methods. Harvard University Press, Cambridge (2014)
20. Preston, S., Heuveline, P., Guillot, M.: Demography: Measuring and Modeling Population Processes, 1st edn. Wiley, Hoboken (2000)
21. Majumder, P., Ghosh, S., Mitra, M.: Ordering results of extreme order statistics from heterogeneous Gompertz–Makeham random variables. Statistics **54**(3), 595–617 (2020). https://doi.org/10.1080/02331888.2020.1750014
22. Rao, B.L.S.P.: On some analogues of lack of memory properties for the Gompertz distribution. Commun. Stat. Theory Methods **47**(18), 4415–4421 (2017). https://doi.org/10.1080/03610926.2017.1376084
23. Kundu, A., Chowdhury, S., Balakrishnan, N.: Ordering properties of the smallest and largest lifetimes in Gompertz-Makeham model. Commun. Stat. Theory Methods **52**(3), 643–669 (2023)
24. Goldstein, J.R., Lee, R.D.: Demographic perspectives on the mortality of COVID-19 and other epidemics. Proc. Natl. Acad. Sci. **117**(36), 22035–22041 (2020)
25. Swanson, D.A., Verdugo, R.R. (Eds.).: Socio-demographic Perspectives on the COVID-19 Pandemic. IAP (2023)
26. Yadav, P.K., Yadav, S.: Subnational estimates of life expectancy at birth in India: evidence from NFHS and SRS data. BMC Public Health **24**(1), 1058 (2024)
27. Saikia, N., Jasilionis, D., Ram, F., Shkolnikov, V.M.: Trends and geographic differentials in mortality under age 60 in India. Popul. Stud. **65**(1), 73–89 (2011)
28. Aalaei, M., Atatalab, F.: The Impact of using Iran life table on the present value of future loss of life insurance portfolio. J. Popul. Assoc. Iran **19**(37), 217–248 (2024)
29. Kirby, J.B., Kaneda, T.: Unhealthy and uninsured: exploring racial differences in health and health insurance coverage using a life table approach. Demography **47**(4), 1035–1051 (2010)
30. Lemaire, J.: Automobile Insurance: Actuarial Models (Vol. 4). Springer Science & Business Media (2013)
31. Singh, A., Shukla, A., Ram, F., Kumar, K.: Trends in inequality in length of life in India: a decomposition analysis by age and causes of death. Genus **73**, 1–16 (2017)
32. Komic, S.R., Walters, K.C., Aderibigbe, F., Rao, A.S.R.S., Stansfield, B.K.: Estimating length of stay for simple gastroschisis. J. Surg. Res. **260**, 122–128 (2021)

Chapter 4
Classical and Newer Measures of Population Stability

In Chap. 1, we studied Lotka's stable population theory along with the associated renewal equation. In Chaps. 1 and 2, we saw Malthus' kind of one population model using a differential equation. In this chapter, we will begin by exploring the Lotka–Volterra system of a two-population model and understanding the stability of such a system. We will provide a general overview of a two-population system of differential equations, followed by a standard three-population model known as the susceptible-infection-recovered (SIR) model proposed by Kermack and McKendrick. We will conduct a stability analysis for the three-population SIR model. Later in the chapter (Sect. 4.4), we will introduce newer definitions of population stability in the context of population momentum, and define metrics related to population replacement, and summarize the associated theories.

The foundational concepts and stability analysis outlined in this chapter can be applied to other differential equation models of higher dimensions and the stochastic modeling framework described in Chap. 6.

4.1 Lotka–Volterra Two-Population (Prey and Predator) Model

Let us consider the following two-population differential equation model with two interacting variables, namely, P_1 (prey population) and P_2 (predator population):

$$\begin{aligned} \frac{dP_1}{dt} &= \Lambda P_1 - \mu_1 P_1 P_2, \\ \frac{dP_2}{dt} &= cP_1 P_2 - \mu_2 P_2. \end{aligned} \tag{4.1}$$

A. S. R. Srinivasa Rao, *Mathematical Demography: Theory and Modeling*,
Mathematical Marvels: Texts and Monographs in the Spirit of CR Rao,
https://doi.org/10.1007/978-981-95-6129-2_4

The model in (4.1) is referred to as Lotka–Volterra two-population prey-predator model. In this model, Λ is the birth rate of the prey population, and μ_1 is the interaction parameter describing an interaction between prey and predator populations. Here μ_1 is treated as an attack rate of predator on prey, and c is a growth parameter contributing to a growth in predator population. The parameter μ_2 is the death rate of the predator population due to a shortage of food (prey). All the four parameters described are positive constants. The quantity Λ is referred to as the intrinsic natural growth of the prey population in the absence of the predator population. The term $\mu_1 P_1 P_2$ in (4.1) represents the loss of prey population and the term $c P_1 P_2$ in (4.1) represents growth in the predator population. The Lotka–Volterra framework described above is a nonlinear system of differential equations and forms an insightful model in population dynamics or ecological dynamics of two populations. Instead of P_1 and P_2, one can consider a time-dependent model of the Lotka–Volterra system with $P_1(t)$ and $P_2(t)$. Let us do a preliminary analysis using the model in (4.1). The Lotka–Volterra framework has been widely applied to understand various ecological and biological dynamics.

4.1.1 Preliminary Analysis

Let us consider the first equation from the model in (4.1), i.e., $\frac{dP_1}{dt} = \Lambda P_1 - \mu_1 P_1 P_2$. We note that $\frac{dP_1}{dt} = 0$ if, and only if, either $P_1 = 0$ or $P_2 = \frac{\Lambda}{\mu_1}$. Because $\Lambda P_1 - \mu_1 P_1 P_2 = 0$, implying

$$P_1 \left(\Lambda - \mu_1 P_2\right) = 0,$$

$$\implies P_1 = 0 \text{ or } P_2 = \frac{\Lambda}{\mu_1}.$$

This preliminary analysis indicates that whenever the predator population density satisfies $P_2 < \frac{\Lambda}{\mu_1}$, it leads the prey population to grow. Whenever $P_2 > \frac{\Lambda}{\mu_1}$, the prey population will decline. Now, let us consider the second equation from the model in (4.1), i.e., $\frac{dP_2}{dt} = cP_1 P_2 - \mu_2 P_2$. We note that $\frac{dP_2}{dt} = 0$ if, and only if, either $P_2 = 0$ or $P_1 = \frac{\mu_2}{c}$. Because $cP_1 P_2 - \mu_2 P_2 = 0$, implying

$$P_2 \left(cP_1 - \mu_2\right) = 0,$$

$$\implies P_2 = 0 \text{ or } P_1 = \frac{\mu_2}{c}.$$

This concludes that whenever prey population density satisfies $P_1 < \frac{\mu_2}{c}$, then the predator population will decline, and if $P_1 > \frac{\mu_2}{c}$, then the predator population

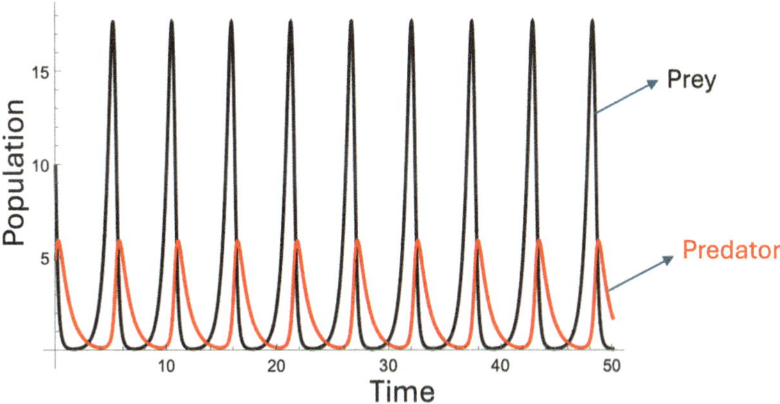

Fig. 4.1 Two-population prey–predator Lotka–Volterra dynamics. The values considered to obtain this graph are $P_1(0)$=10, $P_2(0) = 5$, $\Lambda = 2.1$, $\mu_1 = 1.2$, $c = 1.1$, and $\mu_2 = 0.3$. We notice that the predator population grows when the prey population is high (due to the abundance of food), and the predator population is behind the prey population. When there is not enough prey population, the predator population declines due to a shortage of food

grows. Therefore, the equilibrium points[1] $\left(P_1^*, P_2^*\right)$ for the model in (4.1) are (0, 0) and $\left(\frac{\mu_2}{c}, \frac{\Lambda}{\mu_1}\right)$.

Through this preliminary analysis, we can conclude that both prey and predator must sustain the presence of their respective populations. See Fig. 4.1. The two equilibrium points (0, 0) and $\left(\frac{\mu_2}{c}, \frac{\Lambda}{\mu_1}\right)$ obtained above are used in the Jacobian matrix in the next section to study the stability of the system.

4.1.2 *Stability Analysis*

Let us construct the Jacobian matrix J for the model equations in (4.1), where

$$J = \begin{pmatrix} \frac{\partial}{\partial P_1}\left(\Lambda P_1 - \mu_1 P_1 P_2\right) & \frac{\partial}{\partial P_2}\left(\Lambda P_1 - \mu_1 P_1 P_2\right) \\ \frac{\partial}{\partial P_1}\left(c P_1 P_2 - \mu_2 P_2\right) & \frac{\partial}{\partial P_1}\left(c P_1 P_2 - \mu_2 P_2\right) \end{pmatrix}$$

$$= \begin{pmatrix} \Lambda - \mu_1 P_2 & -\mu_1 P_1 \\ c P_2 & c P_1 - \mu_2 \end{pmatrix}. \tag{4.2}$$

[1] A model of the form $\frac{dP}{dt} = f(P, t)$ has a steady-state solution or equilibrium solution, say, P^*, where $f(P^*, t) = 0$ for all values of t. At steady state the time becomes invariant. The point P^* is referred as asymptotically stable if all local trajectories eventually (i.e., as $t \to \infty$) approach toward the equilibrium. If all trajectories move away from the equilibrium, then the equilibrium becomes unstable. We will see steady-state solutions in stochastic process models as well, for instance, refer to Chap. 6. The steady-state solutions are also called critical points.

See Appendix I for the definition and basics of the Jacobian matrix and eigenvalues. The Jacobian matrix in (4.2) evaluated at the equilibrium point $\left(P_1^*, P_2^*\right)^T = (0,0)^T$ is represented by

$$J|_{(0,0)^T} = \begin{pmatrix} \Lambda - \mu_1 P_2 & -\mu_1 P_1 \\ cP_2 & cP_1 - \mu_2 \end{pmatrix}\Bigg|_{(0,0)^T} = \begin{pmatrix} \Lambda & 0 \\ 0 & -\mu_2 \end{pmatrix}. \tag{4.3}$$

Let us find two eigenvalues, say λ_1, λ_2 for the 2×2 matrix $J|_{(0,0)^T}$ in (4.3). These are found by obtaining the determinant of $\left[J|_{(0,0)^T} - \lambda I\right]$, i.e., $\det\left[J|_{(0,0)^T} - \lambda.I\right]$, where $\lambda = \begin{pmatrix} \lambda_1 \\ \lambda_2 \end{pmatrix}$, and the 2×2 identity matrix $I = \begin{pmatrix} 1 & 0 \\ 0 & 1 \end{pmatrix}$, and solving for $\det\left[J|_{(0,0)^T} - \lambda.I\right] = 0$. The equation $\det[J - \lambda.I] = 0$ is also referred to as the characteristic equation. The characteristic equation for $J|_{(0,0)^T}$ is

$$\begin{aligned} \det\left[J|_{(0,0)^T} - \lambda.I\right] = 0 &\implies \begin{vmatrix} \Lambda - \lambda_1 & 0 \\ 0 & -\mu_2 - \lambda_2 \end{vmatrix} = 0, \\ &\implies (\Lambda - \lambda_1)(-\mu_2 - \lambda_2) = 0, \\ &\implies \lambda_1 = \Lambda \text{ and } \lambda_2 = -\mu_2. \end{aligned} \tag{4.4}$$

Since $\Lambda > 0$ and $\mu_2 > 0$ in (4.1), one eigenvalue of $J|_{(0,0)^T}$ is Λ (we know that by assumption $\Lambda > 0$), and the second eigenvalue of $J|_{(0,0)^T}$ is $-\mu_2$ (we know $\mu_2 > 0$). Therefore, $(0, 0)$ is a saddle point.[2] This indicates that the model at equilibrium point $(0, 0)$ is unstable. For an equilibrium to be stable, both eigenvalues at that point must be negative. Using $P_1 = \frac{\mu_2}{c}$ and $P_2 = \frac{\Lambda}{\mu_1}$ that we saw in Sect. 4.1.1, let us construct the Jacobian matrix at the equilibrium point $\left(P_1^*, P_2^*\right)^T = \left(\frac{\mu_2}{c}, \frac{\Lambda}{\mu_1}\right)^T$. That is,

$$J|_{\left(\frac{\mu_2}{c}, \frac{\Lambda}{\mu_1}\right)^T} = \begin{pmatrix} \Lambda - \mu_1 P_2 & -\mu_1 P_1 \\ cP_2 & cP_1 - \mu_2 \end{pmatrix}\Bigg|_{\left(\frac{\mu_2}{c}, \frac{\Lambda}{\mu_1}\right)^T} = \begin{pmatrix} 0 & -\frac{\mu_1 \mu_2}{c} \\ \frac{c\Lambda}{\mu_1} & 0 \end{pmatrix}. \tag{4.5}$$

[2] The *critical points* or *steady-state solutions* or equilibrium solutions of a linear system of equations with two variables are stable, if, and only of, both roots of the characteristic equation have nonpositive real roots and are asymptotically stable if, and only if, both roots have negative real parts. The critical point of a two-variable linear system is called a *node* if λ_1, λ_2 of the corresponding characteristic equation are real, distinct, same sign. A critical point of a two-variable linear system is called a *saddle point* if λ_1, λ_2 are real, distinct, and have opposite sign. A critical point of a two-variable linear system is a *focus* if λ_1, λ_2 are complex conjugates, but not pure imaginary. If the critical points of a linearized system are stable, then the critical points of the corresponding nonlinear system are also stable. These facts also hold true for linear systems with more than two variables. More discussion on these aspects is in Appendix I.

Let us again find two eigenvalues, λ_1, λ_2, for the 2×2 matrix $J|_{\left(\frac{\mu_2}{c}, \frac{\Lambda}{\mu_1}\right)^T}$ in (4.5). The characteristic equation for $J|_{\left(\frac{\mu_2}{c}, \frac{\Lambda}{\mu_1}\right)^T}$ is

$$\det\left[J|_{\left(\frac{\mu_2}{c}, \frac{\Lambda}{\mu_1}\right)^T} - \lambda.I\right] = 0 \Longrightarrow \begin{vmatrix} -\lambda_1 & -\frac{\mu_1\mu_2}{c} \\ \frac{c\Lambda}{\mu_1} & -\lambda_2 \end{vmatrix} = 0,$$

$$\Longrightarrow \lambda_1\lambda_2 + \Lambda\mu_2 = 0,$$

$$\Longrightarrow \lambda_1 = +i\sqrt{\Lambda\mu_2} \text{ and } \lambda_2 = -i\sqrt{\Lambda\mu_2}. \tag{4.6}$$

The eigenvalues in (4.6) indicate they are purely imaginary and the equilibrium $\left(\frac{\mu_2}{c}, \frac{\Lambda}{\mu_1}\right)^T$ is unstable (not stable). For more on complex numbers, refer to Appendix C, and for details on Jacobian matrix, equilibrium analysis, refer to Appendix I.

4.1.3 Phase-Plane Analysis

A phase-plane analysis is useful in understanding the dynamics of nonlinear dynamical systems, especially the dynamics of two variables at a time. Such an analysis is helpful in graphically depicting limit cycles, equilibrium points, and trajectories. A phase-plane analysis can also be done for higher-dimensional systems. Let us now do the phase-plane analysis of two variables P_1 and P_2 of the prey–predator model in (4.1). To do this, let us first divide $\frac{dP_2}{dt}$ by $\frac{dP_1}{dt}$, i.e., divide the second equation in (4.1) by the first equation in (4.1). This gives us

$$\frac{dP_2}{dP_1} = \frac{cP_1P_2 - \mu_2P_2}{\Lambda P_1 - \mu_1P_1P_2} \tag{4.7}$$

$$= \frac{\mu_2P_2\left(\frac{cP_1}{\mu_2} - 1\right)}{\Lambda P_1\left(1 - \frac{\mu_1P_2}{\Lambda}\right)}. \tag{4.8}$$

A phase portrait can be constructed to understand the vector fields and trajectories and behavior of the model. See Fig. 4.2. Further, let us do a transformation of variables to reduce the original system of two variables $\{P_1, P_2\}$ with four parameters $\{\Lambda, \mu_1, c, \mu_2\}$ into two variables with two parameters as follows:

$$X = \frac{cP_1}{\mu_2}, \text{ and } Y = \frac{\mu_1P_2}{\Lambda}, \tag{4.9}$$

$$\Longrightarrow$$

$$P_1 = \frac{X\mu_2}{c} \text{ and } P_2 = \frac{Y\Lambda}{\mu_1}. \tag{4.10}$$

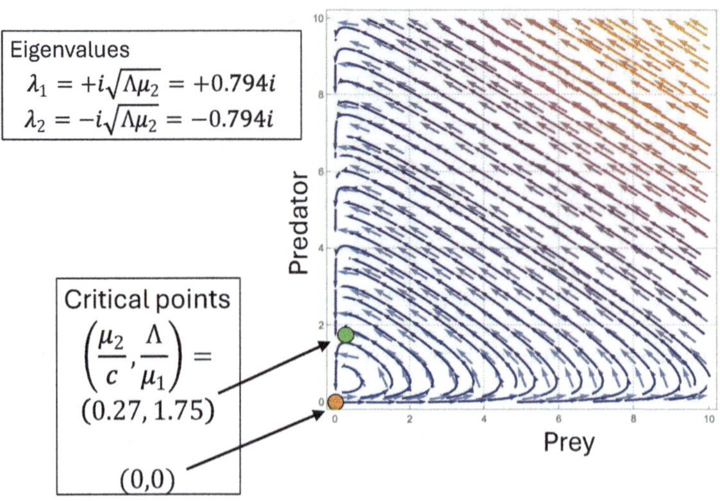

Fig. 4.2 A phase portrait of a two-population prey–predator model. Numerical values used in Fig. 4.1 are considered here to generate the portrait

This implies

$$\frac{dX}{dt} = \frac{c}{\mu_2}\frac{dP_1}{dt},$$
$$\frac{dY}{dt} = \frac{\mu_1}{\Lambda}\frac{dP_2}{dt}. \tag{4.11}$$

Substituting $\frac{dP_1}{dt}$ and $\frac{dP_2}{dt}$ in (4.11), we get

$$\frac{dX}{dt} = \frac{c}{\mu_2}\left(\Lambda P_1 - \mu_1 P_1 P_2\right),$$
$$\frac{dY}{dt} = \frac{\mu_1}{\Lambda}\left(c P_1 P_2 - \mu_2 P_2\right). \tag{4.12}$$

Substituting $P_1 = \frac{X\mu_2}{c}$ and $P_2 = \frac{Y\Lambda}{\mu_1}$ (from (4.10)) in (4.12), we obtain two differential equations with variables X and Y and two parameters Λ and μ_2 as follows:

$$\frac{dX}{dt} = X\Lambda\left(1 - Y\right),$$
$$\frac{dY}{dt} = \mu_2 Y\left(X - 1\right). \tag{4.13}$$

Using the reduced model in (4.13), one can plot X and Y with two parameters Λ and μ_2 to generate another phase portrait (Exercise 4.11).

4.2 A General Two-Population Model: Perturbation Analysis

Let us consider a nonlinear system of two populations P_1 and P_2 as follows:

$$\frac{dP_1}{dt} = f_1(P_1, P_2),$$
$$\frac{dP_2}{dt} = f_2(P_1, P_2), \tag{4.14}$$

with an equilibrium solution (P_1^*, P_2^*). Functions f_1 and f_2 in (4.14) describe the relationship between two variables P_1 and P_2. Let us add small perturbations p_1^* and p_2^* to P_1^* and P_2^*, respectively. Then the system in (4.14) becomes

$$\frac{d\left(P_1 + p_1^*\right)}{dt} = f_1(P_1 + p_1^*, P_2 + p_2^*),$$
$$\frac{d\left(P_2 + p_2^*\right)}{dt} = f_2(P_1 + p_1^*, P_2 + p_2^*). \tag{4.15}$$

By ignoring higher-order Taylor series expansion terms, we write the dynamics of $\frac{dp_1^*}{dt}$ and $\frac{dp_2^*}{dt}$ as follows:

$$\frac{dp_1^*}{dt} = \frac{\partial f_1}{\partial P_1}(P_1^*, P_2^*)p_1^* + \frac{\partial f_1}{\partial P_2}(P_1^*, P_2^*)p_2^*,$$
$$\frac{dp_2^*}{dt} = \frac{\partial f_2}{\partial P_1}(P_1^*, P_2^*)p_1^* + \frac{\partial f_2}{\partial P_2}(P_1^*, P_2^*)p_2^*. \tag{4.16}$$

The system in (4.16) became linear, and the corresponding perturbation analysis conducted is called linear perturbation analysis. Substituting $p_1^* = Pe^{\lambda t}$ and $p_2^* = Qe^{\lambda t}$ in (4.16), we get

$$\lambda Pe^{\lambda t} = \frac{\partial f_1}{\partial P_1}(P_1^*, P_2^*)Pe^{\lambda t} + \frac{\partial f_1}{\partial P_2}(P_1^*, P_2^*)Qe^{\lambda t},$$
$$\lambda Qe^{\lambda t} = \frac{\partial f_2}{\partial P_1}(P_1^*, P_2^*)Pe^{\lambda t} + \frac{\partial f_2}{\partial P_2}(P_1^*, P_2^*)Qe^{\lambda t}. \tag{4.17}$$

Dividing both sides of equations in (4.17) by $e^{\lambda t}$, we get the following $Ax - \lambda I = 0$ type of equation discussed above:

$$\begin{pmatrix} \frac{\partial f_1}{\partial P_1} - \lambda & \frac{\partial f_1}{\partial P_2} \\ \frac{\partial f_2}{\partial P_1} & \frac{\partial f_2}{\partial P_2} - \lambda \end{pmatrix} \begin{pmatrix} P \\ Q \end{pmatrix} = 0, \tag{4.18}$$

$$\Longrightarrow \begin{vmatrix} \frac{\partial f_1}{\partial P_1} - \lambda & \frac{\partial f_1}{\partial P_2} \\ \frac{\partial f_2}{\partial P_1} & \frac{\partial f_2}{\partial P_2} - \lambda \end{vmatrix} = 0. \tag{4.19}$$

From the determinant in (4.19), we obtain the following characteristic equation of the system:

$$\left(\frac{\partial f_1}{\partial P_1} - \lambda\right)\left(\frac{\partial f_2}{\partial P_2} - \lambda\right) = 0$$

$$\Longrightarrow \lambda^2 - \lambda\left(\frac{\partial f_1}{\partial P_1} + \frac{\partial f_2}{\partial P_2}\right) + \left(\frac{\partial f_1}{\partial P_1}\frac{\partial f_2}{\partial P_2} - \frac{\partial f_1}{\partial P_2}\frac{\partial f_2}{\partial P_1}\right) = 0. \tag{4.20}$$

Thus,

$$\lambda_1 = \frac{1}{2}\left[\left(\frac{\partial f_1}{\partial P_1} + \frac{\partial f_2}{\partial P_2}\right) + \sqrt{\left(\frac{\partial f_1}{\partial P_1} + \frac{\partial f_2}{\partial P_2}\right)^2 - 4\left(\frac{\partial f_1}{\partial P_1}\frac{\partial f_2}{\partial P_2} - \frac{\partial f_1}{\partial P_2}\frac{\partial f_2}{\partial P_1}\right)}\right],$$

$$\lambda_2 = \frac{1}{2}\left[\left(\frac{\partial f_1}{\partial P_1} + \frac{\partial f_2}{\partial P_2}\right) - \sqrt{\left(\frac{\partial f_1}{\partial P_1} + \frac{\partial f_2}{\partial P_2}\right)^2 - 4\left(\frac{\partial f_1}{\partial P_1}\frac{\partial f_2}{\partial P_2} - \frac{\partial f_1}{\partial P_2}\frac{\partial f_2}{\partial P_1}\right)}\right].$$

The equilibrium points or the critical points of the system of equations in (4.14) are stable if the values of λ_1 and λ_2 are nonpositive real values. The equilibrium points are asymptotically stable if both characteristic roots or eigenvalues have negative real values, i.e., if $-\left(\frac{\partial f_1}{\partial P_1} + \frac{\partial f_2}{\partial P_2}\right) > 0$ and $\left(\frac{\partial f_1}{\partial P_1}\frac{\partial f_2}{\partial P_2} - \frac{\partial f_1}{\partial P_2}\frac{\partial f_2}{\partial P_1}\right) > 0$. Further discussion on eigenvalues and equilibrium points is in Appendix I.

In this section, a classical two-population prey–predator model is introduced and stability of the system is described. This framework can be extended to higher dimensions with additional variables. The fear factor of prey in the presence of a predator also changes the prey–predator populations and alters the dynamics from the classical dynamics presented above. For example, refer to [1, 2]. There have been other extensions and advancements in Lotka–Volterra-type population models. For more information on prey–predator type of models, refer to [3–21].

4.3 Three-Population Model by Kermack and McKendrick

Let us consider the dynamics of a three-population model consisting of X, susceptible; Y, infected; and Z, recovered individuals and represent it by

$$\frac{dX}{dt} = aZ - bXY,$$
$$\frac{dY}{dt} = bXY - cY,$$
$$\frac{dZ}{dt} = cY - aZ. \tag{4.21}$$

The total population N is represented by $N = X + Y + Z$. All the three parameters $\{a, b, c\}$ in the model in (4.21) are nonnegative real numbers. These parameters are described as follows: a, the rate of removal from Z (and joining the susceptible population X); b, the transmission of infection (or virus) from Y to X; and c, the rate of recovery. The model in (4.21) is referred to as Kermack and McKendrick SIR epidemic model (they both developed it). See references on the SIR model at the end of the section.

4.3.1 *Equilibrium Analysis*

Let us do an equilibrium analysis for the SIR model in (4.21) similar to the analysis done for the Lotka–Volterra two-population model in the previous section. Let (X^*, Y^*, Z^*) be the steady-state or equilibrium solution of the model in (4.21). To find (X^*, Y^*, Z^*), as described previously, we will solve for X, Y, and Z by considering $\left(\frac{dX}{dt} = 0, \frac{dY}{dt} = 0, \frac{dY}{dt} = 0\right)$. The three equations that we obtain to solve for X, Y, and Z are as follows:

$$aZ^* - bX^*Y^* = 0, \tag{4.22}$$
$$bX^*Y^* - cY^* = 0, \tag{4.23}$$
$$cY^* - aZ^* = 0. \tag{4.24}$$

Equation (4.23) implies $Y^* = 0$ or $X^* = \frac{c}{b}$, and Eq. (4.24) implies $Z^* = \frac{cY^*}{a}$. Substituting equilibrium points $X^* = \frac{c}{b}$ and $Z^* = \frac{cY^*}{a}$ in $N = X^* + Y^* + Z^*$, we get

$$N = \frac{c}{b} + Y^* + \frac{cY^*}{a},$$
$$\implies Y^* = \left(N - \frac{c}{b}\right)\left(\frac{a}{a+b}\right). \tag{4.25}$$

Substituting (4.25) in $Z^* = \frac{cY^*}{a}$. we obtain $Z^* = \left(N - \frac{c}{b}\right)\left(\frac{c}{a+b}\right)$. The equilibrium solution or steady-state solution for the model in (4.21) is represented by

$$(X^*, Y^*, Z^*) = \left(\frac{c}{b}, \left(N - \frac{c}{b}\right)\left(\frac{a}{a+b}\right), \left(N - \frac{c}{b}\right)\left(\frac{c}{a+b}\right)\right). \tag{4.26}$$

As $t \rightarrow \infty$, if X, Y, and Z approach the equilibrium in (4.26), we say the equilibrium is stable; else the equilibrium is unstable.

4.3.2 Perturbation Analysis

A perturbation analysis is useful for understanding the behavior of a dynamical system with nonlinear interactions among its variables. In this process, small changes or disturbances are introduced to the system to determine whether there are any resulting changes in its behavior. We will explore the application of this analysis in the SIR model described above.

Since $Z = N - X - Y$, we reduce the three-population model in (4.21) to a two-population infection-susceptible model, and the reduced model is represented by

$$\begin{aligned} \frac{dX}{dt} &= a\,(N - X - Y) - bXY, \\ \frac{dY}{dt} &= bXY - cY. \end{aligned} \tag{4.27}$$

Let us add a small perturbation (x, y) to (X^*, Y^*) (similar to the perturbation analysis discussed in Sect. 4.2). The system of equations (4.27) with perturbation (x, y) becomes

$$\begin{aligned} \frac{dX}{dt} &= \frac{d(X^* + x)}{dt} = a\left(N - (X^* + x) - (Y^* + y)\right) - b(X^* + x)(Y^* + y), \\ \frac{dY}{dt} &= \frac{d(Y^* + x)}{dt} = b(X^* + x)(Y^* + x) - c(Y^* + x). \end{aligned} \tag{4.28}$$

This implies

$$\begin{aligned} \frac{dX^*}{dt} + \frac{dx}{dt} &= aN - a(X^* + Y^*) - a(x + y) - bX^*Y^* - b(X^*y + Y^*x) - bxy, \\ \frac{dY^*}{dt} + \frac{dy}{dt} &= bX^*Y^* + b(X^*y + Y^*x) + bxy - cY^* - cy. \end{aligned} \tag{4.29}$$

At equilibrium, we have noted earlier in the chapter that $\frac{dX^*}{dt} = 0$ and $\frac{dY^*}{dt} = 0$. This implies $aN - a(X^* + Y^*) - bX^*Y^* = 0$ and $bX^*Y^* - cY^* = 0$. The reduced model with perturbations become

$$\frac{dx}{dt} = -a(x + y) - b(X^* y + Y^* x) - bxy,$$
$$\frac{dy}{dt} = b(X^* y + Y^* x) + bxy - cy. \tag{4.30}$$

Usually, the perturbations are small, so we ignore the terms xy from the model, and the model equations in (4.30) becomes

$$\frac{dx}{dt} = -a(x + y) - b(X^* y + Y^* x) = \left(-a - bY^*\right) x + \left(-a - bX^*\right) y \; (= f_1, \text{say})$$
$$\frac{dy}{dt} = b(X^* y + Y^* x) - cy = \left(bY^*\right) x + \left(bX^* - c\right) y \; (= f_2, \text{ say}). \tag{4.31}$$

The Jacobian matrix corresponding to the model equations in (4.31) is represented by

$$J|_{(X^*,Y^*)^T} = \left.\begin{pmatrix} \frac{\partial f_1}{\partial x} & \frac{\partial f_1}{\partial y} \\ \frac{\partial f_2}{\partial x} & \frac{\partial f_2}{\partial y} \end{pmatrix}\right|_{(X^*,Y^*)^T} = \begin{pmatrix} -a - bY^* & -a - bX^* \\ bY^* & bX^* - c \end{pmatrix}. \tag{4.32}$$

Substituting $X^* = \frac{c}{b}$ in (4.32), we get

$$J|_{(X^*,Y^*)^T} = \begin{pmatrix} -a - bY^* & -a - c \\ bY^* & 0 \end{pmatrix}. \tag{4.33}$$

The characteristic equation corresponding to (4.33) is expressed by

$$\det\left(J|_{(X^*,Y^*)^T} - \lambda I\right) = \begin{vmatrix} -a - bY^* - \lambda & -a - c \\ bY^* & -\lambda \end{vmatrix} = 0 \tag{4.34}$$

$\Longrightarrow$

$$\lambda^2 + \lambda\left(a + bY^*\right) + (a + c)\, bY^* = 0,$$

and

$$\lambda_1 = \frac{-\left(a + bY^*\right) + \sqrt{\left(a + bY^*\right)^2 - 4\,(a + c)\, b}}{2},$$
$$\lambda_2 = \frac{-\left(a + bY^*\right) - \sqrt{\left(a + bY^*\right)^2 - 4\,(a + c)\, b}}{2}. \tag{4.35}$$

Let us now understand the status of the stability at equilibrium or critical points or steady-state solutions based on λ_1 and λ_2 expressed in (4.35). Note that $(a + bY^*)^2 - 4(a + c)b < (a + bY^*)^2$. So, irrespective of the value of $(a + bY^*)^2 - 4(a + c)b$, we will have $\lambda_1 < 0$. Let us now study whether $\lambda_2 < 0$, or $\lambda_2 = 0$, or $\lambda_2 > 0$. If $(a + bY^*)^2 - 4(a + c)b < 0$, then λ_2 in (4.35) will have complex conjugates (see Appendix C). The equilibrium point (or equilibrium solution) is a focus. If $(a + bY^*)^2 - 4(a + c)b > 0$, then $\sqrt{(a + bY^*)^2 - 4(a + c)b}$ will have real roots, and λ_2 in (4.35) is negative (i.e., $\lambda_2 < 0$) because $-(a + bY^*) > \sqrt{(a + bY^*)^2 - 4(a + c)b}$.

In conclusion, when $\sqrt{(a + bY^*)^2 - 4(a + c)b} > 0$, both $\lambda_1 < 0$ and $\lambda_2 < 0$, and the equilibrium point is stable, and it is a node. Refer to Footnote 2 for the definition of a node.

4.3.3 Absence of Removal Rate in a Three-Population Model

Let $a = 0$ in (4.21), i.e., there are no individuals from Z joining the susceptible population X. See Fig. 4.3. The model in (4.21) can be written as

$$\begin{aligned} \frac{dX}{dt} &= -bXY, \\ \frac{dY}{dt} &= bXY - cY, \\ \frac{dZ}{dt} &= cY. \end{aligned} \tag{4.36}$$

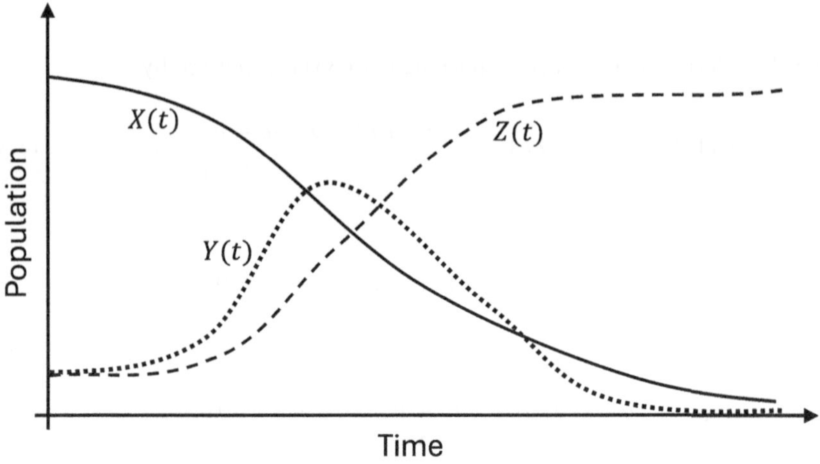

Fig. 4.3 A representation of SIR dynamics. Actual growth and decline of curves depend upon the numerical values of the parameters

For the model in (4.36), let us do a phase-plane analysis of two variables at a time, i.e., we consider (X, Y) and (X, Z). The values of the variables are plotted on coordinates. Dividing $\frac{dX}{dt}$ by $\frac{dZ}{dt}$ from the model equations of (4.36), we get

$$\frac{dX}{dZ} = -\frac{b}{c}X. \tag{4.37}$$

Integrating the Eq. (4.37) through separation of variables, we get

$$\int \frac{1}{X} dX = -\frac{b}{c}\int dZ,$$

$\Longrightarrow$

$$\log X = -\frac{b}{c}Z + C \implies X(t) = X(0)\exp\left[-\frac{b}{c}\left(Z(t) - Z(0)\right)\right]. \tag{4.38}$$

In (4.38), the quantity C is the constant of integration, and $\exp C = X(0)$. Such exponential models are discussed in Chap. 1, Chap. 3, Chap. 8, Appendix H, and other places. The quantity $\frac{b}{c}$ in the model (4.36) is known as the basic reproductive number of the infection. In the absence of recovered individuals joining the susceptible population, we saw that $X(t)$ had become the negative exponential model indicating a decay in the susceptible number over some time $t \in (0, \infty)$. The equation $\frac{dY}{dt} = bXY - cY$ can be rewritten as $\frac{dY}{dt} = (bX - c)\,Y = \left(\frac{b}{c}X - 1\right)cY$ such that the basic reproduction number is situated in $\frac{dY}{dt}$. Dividing $\frac{dY}{dt}$ by $\frac{dX}{dt}$ from the model equations of (4.36), we get

$$\frac{dY}{dX} = -1 + \frac{c}{bX}, \tag{4.39}$$

$\Longrightarrow$

$$Y = -X + \frac{c}{b}\log X + C, \tag{4.40}$$

where C is the constant of integration. Using (4.40), we write the phase-plane function of (X, Y) as $V(X, Y) = Y - X - \frac{c}{b}\log X + C$. The phase portrait can be plotted with actual numerical values of the parameters. See Fig. 4.4.

In this chapter, we have restricted the stability analysis and not included any computations on reproductive rates of infections. However, the net reproductive rates in population dynamics due to Lotka's renewal theory or Euler–Lotka equations were discussed in detail in Chap. 1. The concept of Lotka's population reproductive rates is an inspiration for epidemic models' basic reproductive rates. There are several books and articles available to understand SIR type of models and general mathematical modeling principles. Interested readers may refer to [22–35], and [36–40].

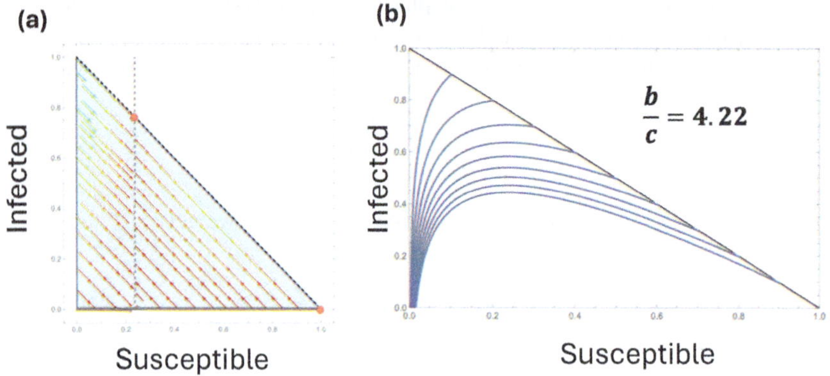

Fig. 4.4 SIR model in (4.36). (**a**) Phase portrait. (**b**) Dynamics and basic reproduction number with $b = 0.713$ and $c = 0.169$

4.4 Recent Advancements in Population Stability

In previous sections of the chapter, we have studied the stability of a system using differential equations. The Lotka–Volterra models provide one of the earliest methods of understanding stability in the context of population biology. In this section, we define and study population stability utilizing a set of births and a set of deaths and without building any differential equation model. This is a new method of assessing population stability, and apart from quantities of set of births and set of deaths, the method also introduces a new metric called population replacement metric.[3] Notably, these newer ideas are not directly linked to Lotka's stable population theory or its integral equation, which were studied earlier in Chap. 1. However, similar to Lotka's original model, this new definition of population stability is also considered in the context of population closed to migration. Population momentum could deter the status of stable population.

Before looking into the technical details of Rao's new approach to population stability, we can summarize the central ideas of this new theory in the following key points:

(i) Population stability is understood through the concept of individuals within a population being elements of sets, with the convergence of these sets being clearly defined.
(ii) Population momentum can cause a stable population to become unstable.
(iii) A population replacement metric has been formulated.

[3] These ideas on population momentum, stability, and replacement metrics, which help determine whether a population in a country or region achieves stability, were presented by Arni S.R.S. Rao at a workshop held in May 2013 at the Mathematical Biosciences Institute in Columbus, Ohio, USA. The workshop was organized to honor the birth centenary of mathematical demographer Nathan Keyfitz. Rao's concepts were later published in 2014 in the *Notices of the American Mathematical Society* [41]. In fact, the 2014 article was dedicated to Alfred J. Lotka.

Recent definitions of population stability introduced utilize a population replacement metric and the mathematical principles of convergence for real numbers, treating individuals within a population as sets of individuals (see Appendix A descriptions related to the convergence of a sequence of real numbers).

4.4.1 Population Replacement Metric and Stability

Let $P_N(t_0)$ represent the set of individuals in a population at time t_0. These individuals can be enumerated as $\{u_1, u_2, \ldots, u_N\}$. Let $|P_N(t_0)|$ represent the cardinality of $P_N(t_0)$ or be the size of the set $P_N(t_0)$. According to the Lyapunov stability, the quantity $|P_N(t_0)|$ is asymptotically stable at a population size U if $||P_N(T)| - U| < \epsilon$ $(\epsilon > 0)$ for some $T > t_0$. A subset P_M of M individuals, drawn from $P_N(t_0)$, is responsible for reproduction within $P_N(t_0)$ during the time interval $[t_0, s]$ for some $s > t_0$. Let $R_\phi(s - t_0)$ represent the reproduction rate of individuals during $[t_0, s]$. Applying $R_\phi(s - t_0)$ among the survivors of $P_N(t_0)$, it results in $|C_n(s - t_0)|$ representing the children born during the interval $[t_0, s]$. This is expressed as

$$R_\phi(s - t_0)\,|P_M(s - t_0)| = |C_n(s - t_0)|\,. \tag{4.41}$$

In (4.41), $|P_M(s - t_0)|$ indicates the effective number of individuals eligible for reproduction during $[t_0, s]$, while $|C_n(s - t_0)|$ denotes the size of the newly born children within the same timeframe.[4] Let $Q_{M_1}(s - t_0)$ be the set of individuals $(M_1 < N)$ who died from the set $P_N(t_0)$ during the interval $[t_0, s]$. Considering the population dynamics over $[t_0, s]$, the size of the population $P(s)$ at time s is represented as follows:

$$P(s) = P_N(t_0) \cup C_n(s - t_0) - Q_{M_1}(s - t_0). \tag{4.42}$$

The size of the set $P(s)$ is $N + n - M_1$, and the individuals in the set $P(s)$ can be enumerated as $\left\{u_1, u_2, \ldots u_{N+n-M_1}\right\}$. The population during $[t_0, s]$ can be termed as stationary if $|C_n(s - t_0)| = \left|Q_{M_1}(s - t_0)\right|$. We can also express $|C_n(s - t_0)| = \left|Q_{M_1}(s - t_0)\right|$ if $|C_n(s - t_0)| \leq \left|Q_{M_1}(s - t_0)\right|$ and $|C_n(s - t_0)| \geq \left|Q_{M_1}(s - t_0)\right|$, which is according to the Cantor–Schröder–Bernstein theorem [42]. Once the equality

$$|C_n(t_0)| = \left|Q_{M_1}(t_0)\right|$$

---

[4] Let A be a set of discrete elements. The notation $|A|$ indicates (in this chapter) the number of elements in the set A or its size. If x is a real number, the notation $|x|$ indicates the absolute value of x.

attains at t_0, a momentum could be generated during $[t_0, s]$ due to one of the inequalities, namely,

$$|C_n(s-t_0)| < \left|Q_{M_1}(s-t_0)\right|,$$

or

$$|C_n(s-t_0)| > \left|Q_{M_1}(s-t_0)\right|.$$

Three following situations arise in the quantities of births and deaths at t_0 and during $[t_0, s]$:

(*i*) $|C_n| > \left|Q_{M_1}\right|$ at t_0 and this inequality may not hold during $[t_0, s]$.
(*ii*) $|C_n| = \left|Q_{M_1}\right|$ at t_0 and this equality may not hold during $[t_0, s]$.
(*iii*) $|C_n| < \left|Q_{M_1}\right|$ at t_0 and this inequality may not hold during $[t_0, s]$. These situations are summarized in Fig. 4.5. The replacement metric d_M with space M_r and population stability introduced in [41] were defined below:

Definition 4.1 (Population Replacement Metric) Let

$$\begin{aligned} S_1 &:\min_s \left| \left\{ \left| |C_n(s-t_0)| - |Q_{M_1}(s-t_0)| \right| : \forall s > t_0 \right\}, \\ S_2 &:\max_s \left| \left\{ \left| |C_n(s-t_0)| - |Q_{M_1}(s-t_0)| \right| : \forall s > t_0 \right\}. \end{aligned} \tag{4.43}$$

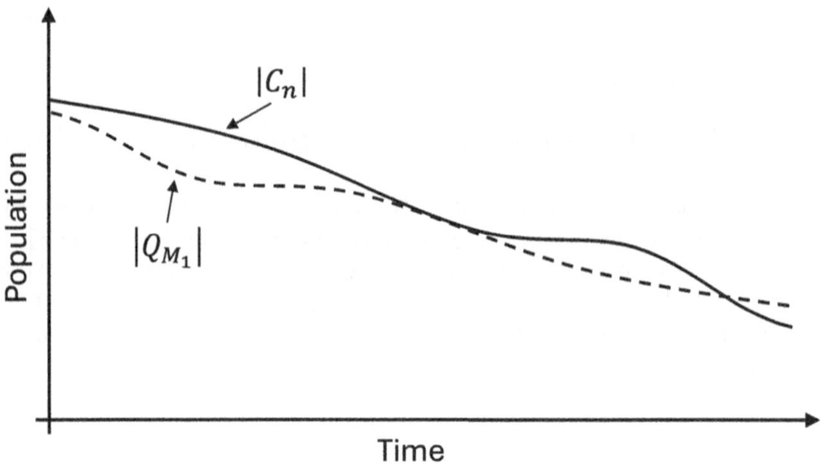

Fig. 4.5 Dynamics of births and deaths over a period of time $[t_0, s]$. The set C_n represents the number of newborn, and the set Q_{M_1} represents the number of deaths out of the set P_N

Let $M_r = [S_1, S_2] \subset \mathbb{R}^+$ and

$$M = \left\{ \left| \min_s \left\{ |C_n(s - t_0)|, \left| Q_{M_1}(s - t_0) \right| \right\} - \max_s \left\{ |C_n(s - t_0)|, \left| Q_{M_1}(s - t_0) \right| \right\} \right| : s > t_0 \right\},$$

with the replacement metric

$$d_M(x, y) = \frac{|x - y|}{k}, \tag{4.44}$$

where $d_M : M \times M \to M_r$, and $[t_0, s]$ is divided into $k + 1$ time intervals.

Remark 4.2 It can be observed that (M, d_M) is a metric space. For more information on metric spaces, please refer to Appendix A. Although d_M is defined in Definition 4.1 as $d_M(x, y) = \frac{|x-y|}{k}$, other similar metrics can be introduced if the difference between two sets $|C_n(s - t_0)|$ and $|Q_{M_1}(s - t_0)|$ appears to be skewed over a time.

Let us further clarify the definition of the population replacement metric in the definition 4.1. Refer to Fig. 4.6. The interval $[t_0, s]$ is divided into the following $k + 1$ time intervals

$$[t_0, s] = [t_0, s_1) \cup [s_1, s_2) \cup \ldots \cup [s_k, s]. \tag{4.45}$$

The quantities S_1 and S_2 are derived from data collected on births and deaths during these $k + 1$ time intervals, i.e.,

$$S_1 = \min \left\{ \left| |C_n(s_1 - t_0)| - |Q_{M_1}(s_1 - t_0)| \right|, \left| |C_n(s_2 - s_1)| - |Q_{M_1}(s_2 - s_1)| \right|, \ldots, \left| |C_n(s - s_k)| - |Q_{M_1}(s - s_k)| \right| \right\}, \tag{4.46}$$

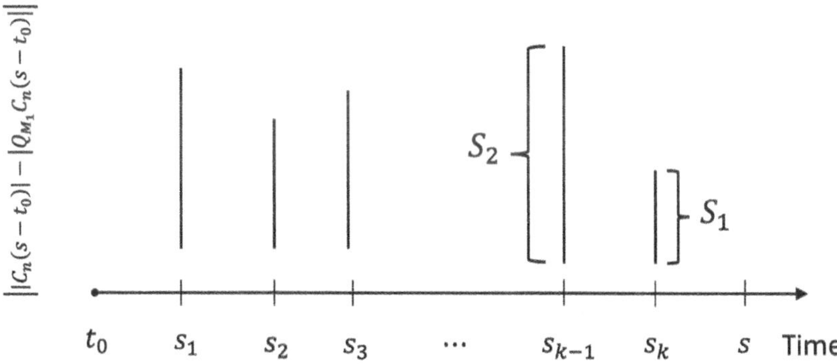

Fig. 4.6 Quantities associated with the population replacement metric defined in 4.1. The horizontal axis displays data obtained from two sets C_n and Q_{M_1}, i.e., $\left| |C_n(s_1 - t_0)| - |Q_{M_1}(s_1 - t_0)| \right|$, ..., $\left| |C_n(s - s_k)| - |Q_{M_1}(s - s_k)| \right|$. The quantities S_1 and S_2 are computed using (4.46) and (4.47)

$$S_2 = \max\left\{\left||C_n(s_1 - t_0)| - |Q_{M_1}(s_1 - t_0)|\right|, \left||C_n(s_2 - s_1)| - |Q_{M_1}(s_2 - s_1)|\right|,\right.$$
$$\left. \ldots, \left||C_n(s - s_k)| - |Q_{M_1}(s - s_k)|\right|\right\}. \tag{4.47}$$

In practice, the quantity M is calculated as $|S_3 - S_4|$, where

$$S_3 = \min\{\min\{|C_n(s_1 - t_0)|, |C_n(s_2 - s_1)|, ...|C_n(s - s_k)|\},$$
$$\min\left\{|Q_{M_1}(s_1 - t_0)|, |Q_{M_1}(s_2 - s_1)|, \ldots, |Q_{M_1}(s - s_k)|\right\}\}, \tag{4.48}$$

$$S_4 = \max\{\max\{|C_n(s_1 - t_0)|, |C_n(s_2 - s_1)|, ...|C_n(s - s_k)|\},$$
$$\max\left\{|Q_{M_1}(s_1 - t_0)|, |Q_{M_1}(s_2 - s_1)|, \ldots, |Q_{M_1}(s - s_k)|\right\}\}. \tag{4.49}$$

Example 4.3 Let us construct a population replacement metric for a population of 11,430,000 at time t_0 during the time interval $[t_0, s]$. During this period, there were 254,000 births and 243,000 deaths. Assume that these births and deaths are distributed across $k + 1$ intervals (4.45). Let $k = 8$. Suppose the values S_1, S_2, S_3, and, S_4 have been calculated from the data over $k + 1$ intervals (using (4.46), (4.47), (4.48), and (4.49), respectively) resulting in the following quantities: 1375, 5500, 14111, and, 13500. The replacement metric for this population is

$$d_M = \frac{|14111 - 13500|}{8} = 76.3.$$

Definition 4.4 (Population Stability) Let $s_1, s_2, ... \in [t_0, \infty)$. Let

$$f_1(s_1) : \left||C_n(s_1)| - |Q_{M_1}(s_1)|\right|,$$

$$f_2(s_2) : \left||C_n(s_2)| - |Q_{M_1}(s_2)|\right|,$$

and so on for $s_1 < s_2 < ....$ We say the population during the time interval $[t_0, \infty)$ is stable if $f_{s_T} \to 0$ for sufficiently large T, and

$$\frac{d}{ds_T}|C_{n_T}(s_T)| = \frac{d}{ds_T}|Q_{M_T}(s_T)| = 0. \tag{4.50}$$

Suppose the individuals of the set P_N are divided into $k-$independent and nonempty subsets $A(1), A(2), \ldots, A(k)$ such that the total number of individuals in all these independent sets equals the size of the set P_N. Consider a family of sets, denoted as F defined on $A(1),\ A(2), \ldots, A(k)$, such that

$$\bigcup_s A(s) = P_N, \tag{4.51}$$

and members of F are disjoint. Now, suppose an arbitrary subset of size k^* is chosen out of the family F that satisfies the condition of equal births and deaths, i.e., $\binom{F}{k^*}$ sets[5] for which births equal deaths and $F - \binom{F}{k^*}$ sets for which births do not equal deaths. Under these conditions, the Theorem 4.5 can be proved.

Theorem 4.5 (Not Stable Criterion) *Let $P_N(t_0)$ be given. Suppose each member of $\binom{F}{k^*}$ satisfies the condition of equal births and deaths, while $F - \binom{F}{k^*}$ does not satisfy the condition of equal births and deaths for $t \geq t_0$. In this case, this need not lead the population to stability.*

Proof When births and deaths are not equal, the degree of the difference between these two quantities at t_0 and at a time $t > t_0$ determines the status of the stability of P_N. The proof uses various possibilities when births and deaths are not equal and develops arguments to show that P_N need not attain stability.

When the numbers of births and deaths are not equal, the degree of difference between these two quantities at time t_0 and at a later time $t > t_0$ determines the stability status of the population P_N. The proof explores various scenarios in which births and deaths are unequal and builds arguments to demonstrate that P_N need not achieve stability: Refer to [41] for a detailed proof.

The methods introduced in this section, which utilize population replacement metrics along with sets of births and deaths, will be beneficial when actual data is available. Understanding whether a population achieved stability in the past can be clarified through these methods. The methods can be helpful in formulating policies for the future population growth. For further details and complete proofs of the theorems, please refer to the source [41]. There have been efforts to understand the impact of population momentum in stationary and stable status of a population [41, 43–51]. For a general discussion on the stability of systems from a dynamical systems perspective, see [52]. For information on Lyapunov methods related to global stability, see [53, 54]. □

4.5 Exercises

Exercise 4.6 Let $P_N(t_0)$ be the set of individuals in a population at time t_0 and $|P_N(t_0)|$ be the size of the set. Let $P_M \subset P_N(t_0)$ be a set of M individuals

[5] The notation $\binom{N}{k}$ is commonly used to indicate that k elements are chosen (selected) out of N elements or k sets out of N sets are chosen. The order of selection does not matter. The formula for $\binom{N}{k}$ is $\frac{N!}{k!(N-k)!}$.

responsible for reproducing and $C_n(s - t_0)$ be the set of children born to P_M during the time interval $[t_0, s]$. Let $Q_{M_1}(s - t_0)$ be the set of deaths from the set $P_N(t_0)$ during the interval $[t_0, s]$, where $M_1 < N$. Given that $P_N(t_0) = 23{,}547{,}349$, $|C_n(s - t_0)| = 671{,}000$, and $Q_{M_1}(s - t_0) = 481{,}200$. It is given that the births and deaths are distributed into 9 time intervals, i.e., $k = 9$. Compute the population replacement metric using $S_1 = 2784$, $S_2 = 12{,}000$, $S_3 = 31{,}000$, $S_4 = 25{,}671$. To recollect,

$$
\begin{aligned}
S_1 =& \min \left\{ \left| |C_n(s_1 - t_0)| - |Q_{M_1}(s_1 - t_0)| \right|, \left| |C_n(s_2 - s_1)| - |Q_{M_1}(s_2 - s_1)| \right|, \right. \\
& \left. \ldots, \left| |C_n(s - s_k)| - |Q_{M_1}(s - s_k)| \right| \right\}, \\
S_2 =& \max \left\{ \left| |C_n(s_1 - t_0)| - |Q_{M_1}(s_1 - t_0)| \right|, \left| |C_n(s_2 - s_1)| - |Q_{M_1}(s_2 - s_1)| \right|, \right. \\
& \left. \ldots, \left| |C_n(s - s_k)| - |Q_{M_1}(s - s_k)| \right| \right\},
\end{aligned}
$$

$$
\begin{aligned}
S_3 =& \min \{ \min \{ |C_n(s_1 - t_0)|, |C_n(s_2 - s_1)|, \ldots |C_n(s - s_k)| \}, \\
& \min \left\{ |Q_{M_1}(s_1 - t_0)|, |Q_{M_1}(s_2 - s_1)|, \ldots, |Q_{M_1}(s - s_k)| \right\} \}, \\
S_4 =& \max \{ \max \{ |C_n(s_1 - t_0)|, |C_n(s_2 - s_1)|, \ldots |C_n(s - s_k)| \}, \\
& \max \left\{ |Q_{M_1}(s_1 - t_0)|, |Q_{M_1}(s_2 - s_1)|, \ldots, |Q_{M_1}(s - s_k)| \right\} \}.
\end{aligned}
$$

Exercise 4.7 Consider the following Lotka–Volterra system of equations with two populations P_1 and P_2:

$$
\begin{aligned}
\frac{dP_1}{dt} =& \Lambda P_1 - \mu_1 P_1 P_2 - dP_1, \\
\frac{dP_2}{dt} =& c P_1 P_2 - \mu_2 P_2,
\end{aligned}
$$

where all the parameters are as per the definitions in (4.1). The only additional parameter added here is the death rate d of the prey population P_1. Find equilibrium solutions and conduct stability analysis for this system of equations.

Exercise 4.8 Given a SIR model of the form with three variables $\frac{dX}{dt} = -bXY$, $\frac{dY}{dt} = bXY - cY$, $\frac{dZ}{dt} = cY$, with a positive set of parameters a, b, c; obtain an expression for the basic reproductive number using the phase-plane relation $V(X(0), Y(0)) = V(X(\infty), Y(\infty))$ and using that expression obtain an expression for $\frac{DZ}{dt}$ (Taylor expansion terms can be ignored after the second term).

Exercise 4.9 Draw phase portraits for the model equations of Exercise 4.8 with a set of positive parameters.

Exercise 4.10 Consider three following scenarios of birth rate and death rate of a population with the mean age at a time t : (i) Birth rate is equal to the death and less than the mean age of the population, (ii) the birth rate is equal to the death and equal to the mean age of the population, (iii) the birth rate is equal to the death rate and is more than the mean age of the population. Build a framework to understand population stability under all three scenarios using the population stability ideas in Sect. 4.4.

Exercise 4.11 Draw a phase portrait of (X, Y) for the reduced Lotka–Volterra model equations:

$$\frac{dX}{dt} = X\Lambda(1 - Y),$$
$$\frac{dY}{dt} = \mu_2 Y(X - 1),$$

where the parameters are positive constants.

References

1. Bhargava, M., Dubey, B., Upadhyay, R.K.: Exploration of prey-taxis and fear-induced turing patterns in ecological networks. Nonlinear Dynam. **113**, 14071–14101 (2025)
2. Dubey, B., Singh, A.: Spatiotemporal dynamics of prey–predator model incorporating Holling-type II functional response with fear and its carryover effects. Chaos Interdiscip. J. Nonlinear Sci. **34**(5), 053108 (2024)
3. Saleem, M.: Predator-prey relationships: indiscriminate predation. J. Math. Biol. **21**(1), 25–34 (1984)
4. Murty, K.N., Rao, D.V.G.: Approximate analytical solutions of general Lotka-Volterra equations. J. Math. Anal. Appl. **122**(2), 582–588 (1987)
5. Samanta, G.P., Chakrabarti, C.G.: Stochastically perturbed Hopf bifurcation in an extended Volterra-Lotka system. Appl. Math. Lett. **2**(2), 163–166 (1989)
6. Kuang, Y., Smith, H.L.: Global stability for infinite delay Lotka-Volterra type systems. J. Differ. Equ. 103(2), 221–246 (1993)
7. Lee, T.J., Kakehashi, M., Rao, A.S.R.S.: Network models in epidemiology. In: Handbook of Statistics, Vol. 44, pp. 235–256. Elsevier (2021)
8. Zeeman, E.C., Zeeman, M.L.: From local to global behavior in competitive Lotka-Volterra systems. Trans. Amer. Math. Soc. **355**(2), 713–734 (2003)
9. Chattopadhyay, J., Chatterjee, S.: Cross diffusional effect in a Lotka-Volterra competitive system. Nonlinear Phenom. Complex Syst. **4**(4), 364–369 (2001)
10. Morita, Y., Tachibana, K.: An entire solution to the Lotka-Volterra competition-diffusion equations. SIAM J. Math. Anal. **40**(6), 2217–2240 (2009)
11. Sakthivel, K., Baranibalan, N., Kim, J.-H., Balachandran, K.: Stability of diffusion coefficients in an inverse problem for the Lotka-Volterra competition system. Acta Appl. Math. **111**(2), 129–147 (2010)
12. Mubayi, A., Castillo-Chavez, C., Chowell, G., Kribs-Zaleta, C., Siddiqui, N.A., Kumar, N., Das, P.: Transmission dynamics and underreporting of Kala-azar in the Indian state of Bihar. J. Theoret. Biol. **262**(1), 177–185 (2010)
13. Srinivasu, P.D.N., Prasad, B.S.R.V.: Role of quantity of additional food to predators as a control in predator-prey systems with relevance to pest management and biological conservation. Bull. Math. Biol. **73**(10), 2249–2276 (2011)
14. Mandal, P.S., Banerjee, M.: Multiplicative-noise can suppress chaotic oscillation in Lotka-Volterra type competitive model. Math. Model. Nat. Phenom. **7**(6), 23–46 (2012)

15. Gupta, R.P., Chandra, P.: Bifurcation analysis of modified Leslie-Gower predator-prey model with Michaelis-Menten type prey harvesting. J. Math. Anal. Appl. **398**(1), 278–295 (2013)
16. Silva-Dias, L., López-Castillo, A.: Spontaneous symmetry breaking of population: stochastic Lotka-Volterra model for competition among two similar preys and predators. Math. Biosci. **300**, 36–46 (2018)
17. Zhang, G.-B., Hao, Y.-C., Shu, Y.-Q.: Bistable wave speed of a three-species Lotka-Volterra competition system with nonlocal dispersal. Discrete Contin. Dyn. Syst. Ser. B **30**(4), 1280–1313 (2025)
18. Tripathi, J.P., Bugalia, S., Jana, D., Gupta, N., Tiwari, V., Li, J., Sun, G.-Q.: Modeling the cost of anti-predator strategy in a predator-prey system: the roles of indirect effect. Math. Methods Appl. Sci. **45**(8), 4365–4396 (2022)
19. Umrao, A.K., Lamba, S., Srivastava, P.K.: Nonlinear dynamics in a fear-driven predator-prey system: bistability, bifurcations, hydra effect, and optimal harvesting. Math. Methods Appl. Sci. **48**(2), 2194–2223 (2025)
20. Liu, W., Xiao, D., Yi, Y.: Relaxation oscillations in a class of predator-prey systems. J. Differ. Equ. **188**(1), 306–331 (2003)
21. Richards, R.L., Drake, J.M., Ezenwa, V.O.: Do predators keep prey healthy or make them sicker? A meta-analysis. Ecol. Lett. **25**(2), 278–294 (2022)
22. Saxena, R., Jadeja, M., Bhateja, V.: Exploring Susceptible-Infectious-Recovered (SIR) Model for Covid-19 Investigation. Springer, Berlin, Heidelberg (2022)
23. Anderson, R.M., May, R.M.: Infectious Diseases of Humans: Dynamics and Control. Oxford University Press (1991)
24. Rao, A.S.R.S., Maini, P.K.: Influencing HIV/AIDS policy in India through mathematical modelling. UK Success Stories in Industrial Mathematics, pp. 257–261. Springer (2016)
25. Rao, A.S.R.S., Thomas, K., Sudhakar, K., Maini, P.K.: HIV/AIDS epidemic in India and predicting the impact of the national response: Mathematical modeling and analysis. Math. Biosci. Eng. **6**(4), 779–813 (2009)
26. Rao, A.S.R.S., Anderson, R.M.: Helminth dynamics: mean number of worms, reproductive rates. In: Handbook of Statistics, vol. 36, pp. 397–404. Elsevier (2017)
27. Brauer, F., Castillo-Chavez, C., Feng, Z.: Mathematical Models in Epidemiology. With a foreword by Simon Levin. Texts in Applied Mathematics, 69. Springer, New York (2019)
28. Dixit, N.M., Perelson, A.S.: Complex patterns of viral load decay under antiretroviral therapy: influence of pharmacokinetics and intracellular delay. J. Theoret. Biol. **226**(1), 95–109 (2004)
29. Naresh, R., Tripathi, A., Tchuenche, J.M., Sharma, D.: Stability analysis of a time delayed SIR epidemic model with nonlinear incidence rate. Comput. Math. Appl. **58**(2), 348–359 (2009)
30. Misra, A.K., Gupta, A., Venturino, E.: Cholera dynamics with bacteriophage infection: a mathematical study. Chaos Solit. Fract. **91**, 610–621 (2016)
31. Rao, A.S.R.S.: Incubation periods under various anti-retroviral therapies in homogeneous mixing and age-structured dynamical models: a theoretical approach. Rocky Mountain J. Math. **45**(3), 973–1031 (2015)
32. Dubey, B., Dubey, P., Dubey, U.S.: Role of media and treatment on an SIR model. Nonlinear Anal. Model. Control **21**(2), 185–200 (2016)
33. Das, T., Srivastava, P.K.: Stability and codimension 2 bifurcations in an SIR model with incubation delay. Nonlinear Anal. Real World Appl. **79**, Paper No. 104127, 25 pp. (2024)
34. Zou, J., Upadhyay, R.K., Pratap, A., Zhang, Z.: Dynamics of a delayed SIR model for the transmission of PRRSV among a swine population. Adv. Differ. Equ. **2020**, Paper No. 351, 30 pp. (2024)
35. Otunuga, O.M.: Global stability for a 2n+ 1 dimensional HIV/AIDS epidemic model with treatments. Math. Biosci. **299**, 138–152 (2018)
36. Kapur, J.N.: Mathematical Modelling, xii+259 pp. Wiley, New York (1988)
37. Krantz, S.G.: Differential equations—a modern approach with wavelets. Textbooks in Mathematics, xiii+467 pp. CRC Press, Boca Raton, FL (2020). ©2020
38. Upadhyay, R.K., Iyengar, S.R.K.: Introduction to Mathematical Modeling and Chaotic Dynamics, xii+351 pp. CRC Press, Boca Raton, FL, (2014)

39. Rao, A.S.R.S., Pyne, S., Rao, C.R.: Disease Modelling and Public Health, Part A. Handbook of Statistics, Vol. 36. Elsevier (2017)
40. Rao, A.S.R.S., Pyne, S., Rao, C.R.: Disease Modelling and Public Health, Part B. Handbook of Statistics, Vol. 36. Elsevier (2017)
41. Rao, A.S.R.S.: Population stability and momentum. Notic. Am. Math. Soc. **61**(9), 1062–1065 (2014). (https://www.ams.org/notices/201409/rnoti-p1062.pdf)
42. Suppes, P.: Axiomatic Set Theory. The University Series in Undergraduate Mathematics. D. Van Nostrand Co. (1960)
43. Krishnamoorthy, S., Potter, R.G., Pickard, D.K. Population momentum: its relation to the moments of replacement level fertility. Math. Biosci. **53**(1–2), 41–51 (1981)
44. Espenshade, T.J., Olgiati, A.S., Levin, S.A.: On nonstable and stable population momentum. Demography **48**, 1581–1599 (2011)
45. Loichinger, E.: Population Momentum. In: Encyclopedia of Gerontology and Population Aging, pp. 3889–3893. Springer International Publishing, Cham (2022)
46. Feichtinger, G., Rau, R., Novák, A.J.: On the momentum of pseudostable populations. Demograpic Res. **52**(Article 15), 445–478 (2025)
47. Whang, C., Choi, S.: Age structure and population momentum in South Korea. Dev. Soc. **44**(2), 345–363 (2015)
48. Li, N., Tuljapurkar, S.: Population momentum for gradual demographic transitions. Popul. Stud. **53**, 255–262 (1999)
49. Li, N.: The momentum of real population under linear fertility transition. Math. Popul. Stud. **13**, 105–116 (2006)
50. Schoen, R., Jonsson, S.H.: Modeling momentum in gradual demographic transitions. Demography **40**(4), 621–635 (2003)
51. Goldstein, J.R.: Population momentum for gradual demographic transitions: An alternative approach. Demography **39**(1), 65–73 (2002)
52. Rao, A.S.R.S., Krantz, S.G.: Dynamical systems: From classical mechanics and astronomy to modern methods. J. Ind. Inst. Sci. **101**(3), 419–429 (2021)
53. Lakshmikantham, V., Liu, X.Z.: Stability Analysis in Terms of Two Measures. World Scientific Publishing, River Edge, NJ (1993)
54. Izobov, N.A.: Lyapunov Exponents and Stability. Stability, Oscillations and Optimization of Systems, xviii+353 pp. Cambridge Scientific Publishers, Cambridge (2012)

Chapter 5
Mathematics of Stationary Populations

This chapter describes traditional and recent developments in stationary populations' theory, models, and methods.

A stationary population is defined as one in which the total population size remains constant over a finite or infinite period of time, and the distribution of the population does not change. In other words, a population is stationary if it has zero population growth and has constant age structure of the population. In Chap. 1, we discussed Euler–Lotka's fundamental integral equation of populations and the concept of intrinsic rates, which introduce the idea of stationary populations. As noted earlier in Chap. 3, the life table is a stationary population model. The notion of stationary populations, along with related mathematical formulations, has been explored by esteemed mathematicians and statisticians for the past four hundred years.

Notable contributors to the understanding of the fundamental mathematical principles behind stationary populations include mathematician Leonhard Euler (1707–1783), statistician John Graunt (1620–1674), and astronomer Edmond Halley (1656–1742). Refer to [1–5]. The concept of stationarity is not new to physics; ancient mathematicians from India and elsewhere, such as Aryabhata in the fifth century CE and Bhaskara in the eleventh century CE, studied the stationary status of planets, celestial properties, retrograde motions, and planetary orbits [6–11].

Population pyramids, like histograms, are used to depict the population frequencies as a bar graph by age groups. Histograms are generally plotted vertically on a graph but pyramids are drawn as horizontal bars, where each bar describes populations within each age group. A comparison of two population age pyramids can be visualized mathematically as a comparison between two 2-dimensional pyramid-shaped geometric objects. Usually, population pyramids are built for comparing male and female age structures of a particular population for gender differentials in patterns of births, deaths, and migrants. Other examples include age or race-specific stratifications of populations, comparing epidemiological events by gender, etc.

A. S. R. Srinivasa Rao, *Mathematical Demography: Theory and Modeling*, Mathematical Marvels: Texts and Monographs in the Spirit of CR Rao,
https://doi.org/10.1007/978-981-95-6129-2_5

Such pyramids are also often built for total population age structures across populations to compare demographic transitions, population aging indicators, stationary indicators, etc. Suppose we randomly draw a very large sample from a population and record the current age and the subsequent duration of life for each individual. Using this information, suppose we construct two pyramids, one from the age frequencies in the sample we had drawn and the other from the measurement of the subsequent duration of each individual life (length of life after cohort formation), and we ask questions, such as what do these pyramids look like? Do they have a similar structure? It is not always straightforward to make such comparisons, but when the populations are stationary then we could use elegant mathematical identities.

In more recent years, several articles authored by Arni S.R.S. Rao, both independently [12–18] and in collaboration with James R. Carey [19–24], have significantly advanced the theory of stationary populations. They have introduced new measures, linked stationary and nonstationary populations, explored oscillatory properties, and developed Rao's partition theorems related to populations [15], Rao-Carey fundamental theorem in stationary populations [19], etc.

In Sects. 5.1, 5.2, and 5.9, we will first examine the geometric properties of stationary populations using population pyramids and mathematical relationships. Later, in Sects. 5.3, 5.5, and 5.6, we will discuss other recent advancements in the study of stationary populations, namely, Rao–Carey fundamental theorem, oscillatory properties, and Rao's partition theorem of populations. Partition theorem ideas for discrete and continuous age intervals are described in other sections in the chapter. Section 5.7 describes newer measures to compute population replacement level (Q_0) which is an alternate to the traditional NRR (R_0) formula (5.43) of Chap. 1.

5.1 Population Pyramids and Stationary Models

In a stationary population, let $X_0, X_1, \ldots, X_\omega$ be the sets of individuals who are captured at the ages $0, 1, \ldots, \omega$, respectively (here ω is the highest age), and let $n_0, n_1, \ldots, n_\omega$ be their corresponding number of captured individuals in each set. The population age pyramid for these captured individuals is not possible to construct if age at capture is unknown. Suppose these captured individuals are followed till death, and we measure the time till death of each of these individuals after they were captured.

Let Y_0 be the set of individuals who live for 0 years (or 0 units of time) after capture and let m_0 be its size. Here the set Y_0 is formed as a random subset of individuals out of $\cup_{i=0}^{\omega} X_i$. Let Y_1 be the set of individuals who will live for 1 year or 1 unit of time after capture which is formed as a random subset out of $\cup_{i=0}^{\omega} X_i$ and so on the set Y_k formed out of $\cup_{i=0}^{\omega} X_i$. Here the size of Y_k is m_k for $k = 1, 2, \ldots, \omega$. As the captured individuals start dying at various times, then the age pyramid can be constructed for the post-captured population based on the time lived after capture (time lived until the death of an individual after capture is the same as time left to live at the time of capture). Observe that

$$X_0 \neq X_1 \neq X_2 \neq \cdots \neq X_\omega,$$
$$Y_0 \neq Y_1 \neq Y_2 \neq \cdots \neq Y_\omega,$$

because if $x \in X_i$, then $x \notin X_j$ for $i \neq j$ and $y \in Y_i$ then $y \notin Y_j$ for $i \neq j$. Age pyramid at capture and age pyramid after capture until death are identical (by Brouard's theorem[1]); hence, we can write

$$\sum_{y=0}^{\omega} m_y = \sum_{x=0}^{\omega} n_x, \tag{5.1}$$

where m_y is the size of the set of individuals who will live for y years or y units of time, and n_x is the size of the captured individuals set of age x. In addition, by Brouard's theorem, for each $a \in [0, \omega)$, we will have $m_a = n_a$, and

$$\frac{m_a}{\sum_{y=0}^{\omega} m_y} = \frac{n_a}{\sum_{x=0}^{\omega} n_x} \forall a \in [0, \omega). \tag{5.2}$$

The L.H.S. of (5.2) is the fraction of individuals who lived $y$years (or y units of time) after capture and the R.H.S. of (5.2) is the fraction of individuals who were captured at age y. Note that m_a is the size of the set Y_a which is formed individuals out of the set $\cup_{i=0}^{\omega} X_i$; see Fig. 5.1. A statement similar to Brouard's theorem was independently observed by an entomologist James R. Carey through his ecological field experiments on the fruit fly population. He states that "*In a stationary population, the fraction of individuals who are captured at age 'x' is equal to the fraction of individuals who will have 'x' years to live*" [20, 26]. This later became Carey's equality [27]. A novel mathematical proof of Carey's equality was proven by Rao, Arni S.R.S. in 2013 during a chance meeting [28, 29], leading to a collaboration with Carey to publish this theorem [19]. See Sect. 5.3 on the Rao–Carey fundamental theorem. This was later labelled as the stationary population identity [20]. There were other discussions and statements associated with life lived and life left among a cohort of populations and life tables, for example, see Sect. 5.4 in the chapter.

Alternatively, when (5.2) is true for a stationary population, then the number of individuals in each age $a \in [0, \omega)$ in a captured population is equal to the number of individuals in the corresponding age $a \in [0, \omega)$. This implies, in a stationary population, the age pyramid of a captured population and age pyramid constructed

[1] Brouard's theorem (Theorem 4.1, page 160, published in 1989 in French [25]) states "*in a stationary population one can count as many individuals aged 'x' years, then individuals with 'x' years to live*" and a Property 4.2.4 (page 161) which states "*in a stationary population the age pyramid and the pyramid of years to live are identical*".

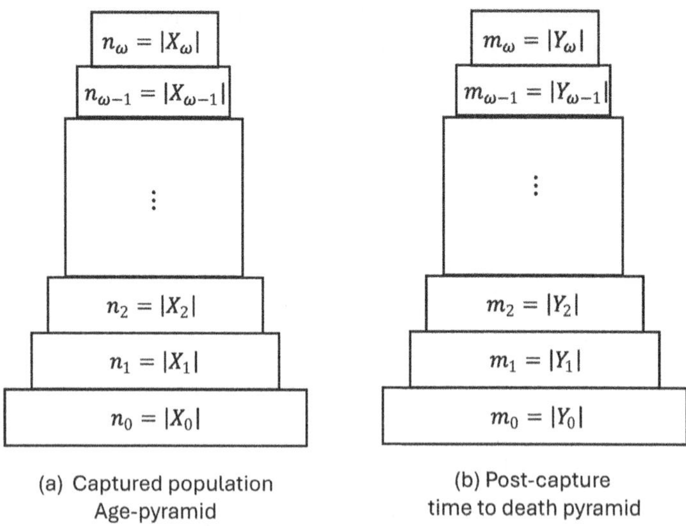

Fig. 5.1 Schematic diagram of (**a**) population age pyramid and (**b**) population pyramid of captured individuals followed until death. The quantities in the rectangles indicate the size of the respective populations within the age of the rectangle

for the captured population until death are identical. It was also demonstrated that life table identities can be used to partition a given large population into stationary and nonstationary components [15, 21].

5.2 Geometry of Stationary Populations

In this section, we describe more geometric properties of stationary populations. See Theorem 5.1 below where each of the rectangle in Fig. 5.1 can be associated to understand the population fractions.

Let $P_i Q_i R_i S_i$ (Fig. 5.1) be the rectangle represented to describe the population who are within the age group $[i, i+1)$ for $i = 0, 1, 2, \ldots, \omega$. Let the area of this rectangle be denoted by $P_i Q_i . P_i S_i$, where $P_i Q_i .$ is the length and $P_i S_i$ is the height of the rectangle as shown in Fig. 5.2. Suppose the height of the rectangle indicates count of one individual and the length of this rectangle indicates the number of such individuals who are in the age group $[i, i+1)$ for $i = 0, 1, 2, \ldots, \omega$, so that $P_i Q_i . P_i S_i$ is the total number of individuals who are in the age group $[i, i+1)$.

Theorem 5.1 *Area of the rectangle for the age group* $[0, 1)$, *i.e.,* $P_0 Q_0 . P_0 S_0$ *is equal to the sum of all the areas lost during the intervals* $[i, i+1)$ *for* $i = 0, 1, 2, \ldots, \omega$ *after each of the rectangle was formed.*

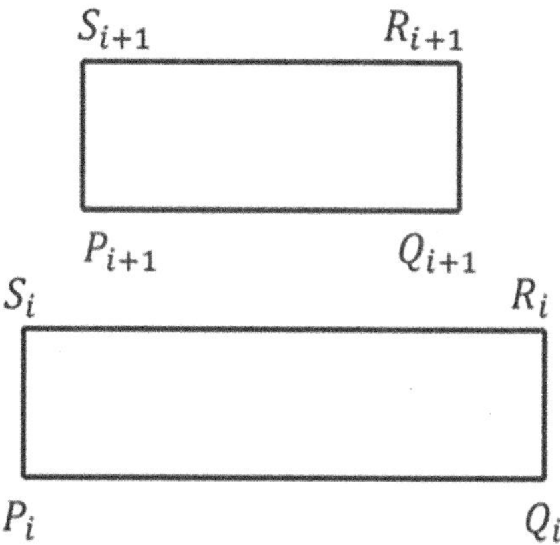

Fig. 5.2 Rectangles within an age pyramid for $i = 0, 1, 2, \ldots, \omega$

Proof Let $u(a, t)$ be the number of individuals who are at age a at time t and let $\int_0^1 u(a, t)da$ be the total number of individuals[2] who are in the age group $[0, 1)$. Let $\int_i^{i+1} u(a, t+i)da$ be the total number of individuals who are in the age group $[i, i+1)$ at time $t+i$ out of the survivors of the cohort $\int_0^1 u(a, t)da$ for $i = 1, 2, \ldots, \omega$. Due to death rates in each age group $[i, i+1)$, there will be a lesser number of individuals in the age group $[i, 1+1)$ than the number of individuals in $[i, i+1)$ for $i = 0, 1, 2, \ldots, \omega$ such that the inequality (5.3) is satisfied.

$$\int_{i-1}^{i} u(a, t+i-1)da > \int_i^{i+1} u(a, t+i)da \text{ for } i = 1, 2, \ldots, \omega. \tag{5.3}$$

The fraction of individuals surviving at $t+i$ out of $\int_0^1 u(a, t)da$ is given by

$$\frac{\int_i^{i+1} u(a, t+i)da}{\int_0^1 u(a, t)da} = \frac{P_i Q_i}{P_0 Q_0}, \tag{5.4}$$

since $P_0 S_0 = P_i S_i$ and $P_i Q_i < P_0 Q_0$. Hence, the area lost due to deaths during $[0, i)$ is

$$(P_0 Q_0 - P_i Q_i)\, P_i S_i = \int_0^1 u(a, t)da - \int_i^{i+1} u(a, t+i)da, \tag{5.5}$$

and the area lost due to deaths during $[i, i+1)$ is

$$(P_i Q_i - P_{i+1} Q_{i+1})\, P_{i+1} S_{i+1} = \int_i^{i+1} u(a, t+i)da - \int_{i+1}^{i+2} u(a, t+i+1)da. \tag{5.6}$$

[2] We request the readers to not be confused with the notation $u(a, t)$ used in Chap. 1, which was designated there for a different purpose while studying Lotka's stable population theory.

From (5.5) and (5.6), we will deduce

$$\int_0^1 u(a,t)da = \sum_{i=0}^{\omega}\left[\int_i^{i+1} u(a,t+i)da - \int_{i+1}^{i+2} u(a,t+i+1)da.\right].$$

Hence, the statement is proved. □

5.3 Rao–Carey Fundamental Theorem in Stationary Populations

The stationary population identity that in a life table or in a stationary population model, the proportion of individuals at age a is equal to the proportion of individuals who have a years to live was described in the introduction of Chap. 5.[3] Below we will state the theorem and provide an outline of the proof of [19]. Further explanations of the proof in [19] are also provided in Sect. 5.3.1.

Theorem 5.2 (Rao and Carey, 2015 [19]) *In a stationary population, let* (X, Y, Z) *be a triplet of column vectors, where* $X = [x_1, x_2, \ldots x_k]^T$ *has information on age at capture of k individuals,* $Y = [y_1, y_2, \ldots, y_k]^T$ *has information on follow-up durations of these k individuals until death, and* $Z = [z_1, z_2, \ldots, z_k]^T$ *has information on length of lives of k individuals, respectively. Let* G_1 *be the graph connecting the coordinates* $\{(1, y_1), (2, y_2), \ldots, (k, y_k)\}$ *on the first quadrant, and let* G_2 *be the graph connecting the coordinates* $\{(-1, x_1), (-2, x_2), \ldots, (-k, x_k)\}$ *on the second quadrant. Let* $\mathcal{H}$ *be a family of graphs consisting of* $k!$ *permutations of graphs on the second quadrant with all possible permutations of the data in X. Then, one of the members of the family* $\mathcal{H}$*, say,* H_g *is a vertical mirror image of* G_1.

[3] In 2013, during a workshop in Columbus, Ohio, USA, honoring the birth centenary of mathematical demographer Nathan Keyfitz, an entomologist from University of California, Davis, James R. Carey presented his talk on experimental and simulated results related to what was then known as Carey's equality. Arni S.R.S. Rao, attending the session, encountered the result for the first time. Rao restated the result in a different way and, within just 45 minutes, proved it using tools from algebra, combinatorics, and analysis. Rao shared his proof to Carey while Carey was still at the podium. This news about the new proof was later published by *Math Digest*, a magazine of the American Mathematical Society [28]. Later, Rao and Carey worked together to publish an article based on the above mathematical proof along with a few more results in the *Journal of Mathematical Biology* (JOMB) in 2015. A rigorous proof involving the geometry of capture age structure of the population and their follow-up durations in an innovative approach received attention among the researchers working in formal demography (see, for example, [29]). One of the two referees who reviewed the above manuscript for JOMB wrote "*...In a very original and unusual way the authors 'trace' the formation of this equality ' understanding the role of each captive subject, and their corresponding follow-up duration in a stationary population. This is indeed very interesting and unusually innovative....*" Another referee wrote "*...It's hard to write down such a proof any more simply than as given here. This proof should also be really useful in making extensions to other captive cohort situations....*"

Proof Because the population is stationary, the equality

$$f_1(a) = f_2(a) \tag{5.7}$$

holds, where f_1 is a probability density function describing capture ages and f_2 is a probability density function describing follow-up times to death. See Appendix B for definition and details on probability density functions. In a large population, assuming the sum of the capture ages of all individuals in the population and their follow-up durations till their deaths is equal to the total length of lives of all individuals, we can write

$$\int_0^\infty x(s)ds + \int_0^\infty y(s)ds = \int_0^\infty z(s)ds, \tag{5.8}$$

where $x \in X$, $y \in Y$, and $z \in Z$. We note here that for the ith individual in the above population, $x_i + y_i = z_i$. In case we treat the age of the captured cohort as discrete, then, instead of considering a probability density function, one can consider probability mass function. Two criteria U_1 and U_2 are defined as follows on individuals in X and Y:

$$\begin{aligned} U_1 &: \left\{x_i \neq x_j \text{ and } y_i \neq y_j \text{ for } i \neq j\right\}, \\ U_2 &: \left\{x_i = x_j \text{ and } y_i = y_j \text{ for at least one } i \neq j\right\}. \end{aligned} \tag{5.9}$$

Suppose U_1 holds. Let us first construct G_1 using an ordered pairs of set S given by

$$S = \left\{(t_1, y_{t_1}), (t_2, y_{t_2}), \ldots, (t_k, y_{t_k})\right\}, \tag{5.10}$$

where $t_1, t_2, \ldots, t_k$ are the k distinct individuals of original population of size k. In (5.10), $y_{t_1}, y_{t_2}, \ldots y_{t_k}$ are defined as follows:

$$\begin{aligned} y_{t_1} &: \max\{y_i : y_i \in Y\}, \\ y_{t_2} &: \max\left\{y_i : y_i \in Y \text{ and } y_i \neq y_{t_1}\right\}, \\ y_{t_3} &: \max\left\{y_i : y_i \in Y \text{ and } y_i \neq y_{t_1}, y_{t_2}\right\}, \\ &\vdots \qquad \vdots \\ y_{t_k} &: \max\left\{y_i : y_i \in Y \text{ and } y_i \neq y_{t_1}, y_{t_2}, \ldots, y_{t_{k-1}}\right\}. \end{aligned} \tag{5.11}$$

The graph G_1 is constructed by joining the coordinate points in S. When S satisfies criterion U_1, then G_1 is a decreasing graph, and if S satisfies U_2, then G_2 is a combination of decreasing and nondecreasing function. There are multiple ways one can construct a corresponding H_g that is a mirror image of G_1. One possibility is to verify if

$$
\begin{aligned}
|t_1 - (-1)| &= 0 \text{ and } \left|y_{t_1} - (x_1)\right| = 0, \\
|t_2 - (-2)| &= 0 \text{ and } \left|y_{t_2} - (x_2)\right| = 0, \\
&\vdots \\
|t_k - (-k)| &= 0 \text{ and } \left|y_{t_k} - (x_k)\right| = 0,
\end{aligned} \tag{5.12}
$$

and, if not, check for other permutation of integers of $\{1, 2, \ldots, k\}$ to subtract from $t_1, t_2, \ldots, t_k$ and subtract corresponding values of X from $\{y_{t_1}, y_{t_2}, \ldots, y_{t_k}\}$ until the absolute values of both the equalities are zeros. By joining the corresponding coordinates, we construct a graph on the second quadrant that is a mirror image of G_1. Let us call the mirror image graph by H_g. Such an equality exists because of the stationary status of the population; there exists an age in X that equal to y_{t_1}, and so on there exists an age in X that is equal to y_{t_k}. That is, through this approach, we are constructing a family of graphs $\mathcal{H}$ in which H_g is identified by verifying if the identity exists. Another approach is as follows: Since the population is stationary, we can construct a set T where

$$
T = \left\{(u_1, x_{u_1}), (u_2, x_{u_2}), \ldots, (u_k, x_{u_k})\right\}, \tag{5.13}
$$

and

$$
\begin{aligned}
x_{u_1} &: \max\{x_i : x_i \in X\}, \\
x_{u_2} &: \max\left\{x_i : x_i \in X \text{ and } x_i \neq x_{u_1}\right\}, \\
x_{t_3} &: \max\left\{x_i : x_i \in X \text{ and } x_i \neq x_{u_1}, x_{u_2}\right\}, \\
&\vdots \qquad \vdots \\
x_{t_k} &: \max\left\{x_i : x_i \in X \text{ and } x_i \neq x_{u_1}, x_{u_2}, \ldots, x_{u_{k-1}}\right\}.
\end{aligned} \tag{5.14}
$$

The graph constructed through the coordinates of T will be a mirror image of G_1 because

$$
\begin{aligned}
|t_1 - (-1)| &= 0 \text{ and } \left|y_{t_1} - (x_{u_1})\right| = 0, \\
|t_2 - (-2)| &= 0 \text{ and } \left|y_{t_2} - (x_{u_2})\right| = 0, \\
&\vdots \\
|t_k - (-k)| &= 0 \text{ and } \left|y_{t_k} - (x_{u_k})\right| = 0.
\end{aligned} \tag{5.15}
$$

Similarly, one can see that H_g exists under the criterion U_2. □

Remark 5.3 (Novelty and Innovation in Rao and Carey, 2015 Theorem [19]) The idea of comparing the age-structure-captured cohort with the individuals' follow-up durations presented in the statement and proof is novel. The proof does not use the concepts of renewal theory discussed in Chap. 1. The other innovation was constructing graphs based on the age structure of the captured data and the follow-up data and matching the images of the graphs. Subsequent researchers also used such intuitions in explaining data on life lived versus life left.

5.3.1 Clarifications on the Proof of the Rao–Carey Theorem

Interpretation of what authors mean by the word "infinite" population and existence of a function H_g in the proof of Theorem 1 in [19] are provided.

Such clarifications are not required in general, but some of the practical demographic readers of the article [19] misunderstood the analytical arguments provided and some readers have wrongly understood the mathematical meaning of the arguments [30, 31]. The proof of the main theorem is novel and provides new insights to visualization of the data arising from demographic cohorts, with no errors (not even minor) in the proof.

In the proof of the Rao–Carey Theorem 1 (page 585–587 of [19]), although they mention the phrase "infinite population," what they meant was "infinitely large discretely measured" or simply "large" population. In fact, in their proof, Rao–Carey used discrete populations to demonstrate that two graphs G_1 and H_g are mirror images to one another.

Another very subtle issue we wish to clarify to our readers is related to graph H_g within the family of graphs $\mathcal{H}$ introduced at the end of the proof. Corresponding to each value of $y_{t_i} \in S$, there exists $x_l \in I_2$, i.e., there exists $x_l \in I_2$ such that $\left|y_{t_i} - x_l\right| = 0$ for some l ($i = l$ or $i \neq l$). Out of $k!$ permutations, there exists only one such permutation, where

$$\left\{\left|y_{t_1} - x_1^{(g)}\right| = 0 \text{ and } \left|y_{t_2} - x_2^{(g)}\right| = 0, \text{ and}\right.$$
$$\left.\left|y_{t_3} - x_3^{(g)}\right| = 0 \cdots \left|y_{t_k} - x_k^{(g)}\right| = 0\right\}$$

for

$$x_1^{(g)}, x_2^{(g)}, x_3^{(g)}, \cdots, x_k^{(g)} \in I_2.$$

The age structure $\left\{x_1^{(g)}, x_2^{(g)}, x_3^{(g)}, \cdots, x_k^{(g)}\right\}$ exists. However, t_1 may not be equal to 1, t_2 may not be equal to 2, and so on. Hence, shifting the indices of $x_1^{(g)}$ to t_1

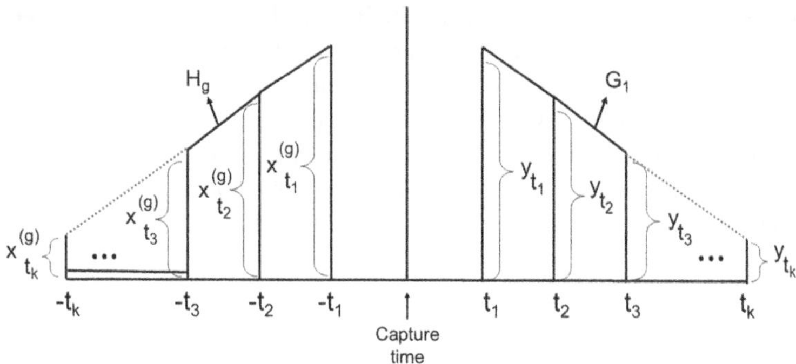

Fig. 5.3 Visualization of the functions H_g and G_1

(if $t_1 \neq 1$), $x_2^{(g)}$ to t_2 (if $t_2 \neq 2$), and so on until shifting the index of $x_k^{(g)}$ to t_k (if $t_k \neq k$) and joining the coordinates

$$S' = \left\{ \left(-t_1, x_{t_1}^{(g)}\right), \left(-t_2, x_{t_2}^{(g)}\right), \cdots, \left(-t_k, x_{t_k}^{(g)}\right) \right\}, \tag{5.16}$$

we will see G_1 is a mirror image of H_g. Here G_1 is constructed using the coordinates in the set S defined in [19] and H_g is constructed using the coordinates in the set S' shown in Fig. 5.3.

By the construction of $S = \{(t_1, y_{t_1}), \ldots, (t_k, y_{t_k})\}$, G_1 can never be a graph of increasing function. Based on U_1 and U_2 in page 585 in [19], they mean that there will be a unique value of the duration of life left corresponding to each captured subject and one of the values of the remaining life left is identical to exactly one of the captured ages. Therefore, H_g exists.

5.3.2 *Computation of Life Expectancy of a Capture Cohort*

Let $c(k, x)$ represent the kth individual captured at age x for $k = 1, 2, \ldots$. The mean age at capture for the total captured population, say $c(0)$, is

$$c(0) = \frac{\int_0^\infty \left(\sum_{k=1}^\infty x c(k, x)\right) dx}{\int_0^\infty \left(\sum_{k=1}^\infty c(k, x)\right) dx}, \tag{5.17}$$

where $\sum_{k=1}^\infty c(k, x)$ represents the number of captured at age x, and

$$\int_0^\infty \left(\sum_{k=1}^\infty c(k, x) \right) dx \tag{5.18}$$

represents the total captured. The number of deaths in the age interval $(0, s)$ after being captured is

$$\int_0^s d(y)dy, \tag{5.19}$$

where $d(y)$ is the number of deaths at age y. The number of captured individuals surviving at age s can be computed using total captured in (5.18) and (5.19) as

$$\int_0^\infty \left(\sum_{k=1}^{\infty} c(k,x)\right) dx - \int_0^s d(y)dy. \tag{5.20}$$

Similarly, the number of captured individuals surviving at age $(n+1)s$ can be computed using

$$\int_0^\infty \left(\sum_{k=1}^{\infty} c(k,x)\right) dx - \sum_{n=0}^{\infty}\left(\int_{ns}^{(n+1)s} d(y)dy\right). \tag{5.21}$$

The life expectancy of the capture cohort, say $E[c(0)]$, is computed using (5.18) and (5.21) as

$$\frac{\int_0^\infty \int_0^\infty \left(\sum_{k=1}^{\infty} c(k,x)\right) dx - \int_0^\infty \left[\sum_{n=0}^{\infty}\left(\int_{ns}^{(n+1)s} d(y)dy\right)\right] ds}{\int_0^\infty \left(\sum_{k=1}^{\infty} c(k,x)\right) dx}. \tag{5.22}$$

The result (5.22) was derived in [19].

5.4 Historical Perspectives on Life Lived and Life Left Data

In a life table, the total births are equal to the total deaths, and the birth rate of a life table and the death rate of a life table are identical. Based on this important feature of a life table, several identities were evolved in the theory of stationary population. We have already mentioned in this chapter a few results on life lived and life left, namely, Brouard's theorem on life table identities, Carey's equality of life table populations, and Rao–Carey fundamental theorem in stationary populations. In 1989, Kim Y.J. and Aron J.L. [32] came up with an identity inspired by renewal theory in probability, and their main statement was as follows: In a stationary population, the average age, say A, is equal to the average age of expected remaining life, say E, i.e., $A = E$, where

$$A = \frac{\int_0^\infty a p_a da}{\int_0^\infty p_a da}, \tag{5.23}$$

and

$$E = \frac{\int_0^\infty e_a^0 p_a da}{\int_0^\infty p_a da}, \tag{5.24}$$

where $p_a = \exp\left(-\int_0^a \mu_s ds\right)$ is the survival probability function to live up to age a and it can be seen in (3.16). Using (3.11), we note that $e_a^0 = \int_0^\omega \left(\frac{1}{l_a}\int_0^\infty l_{a+t}dt\right)$. We will see in Chap. 8 that the function μ is treated as hazard rate in survival analysis. Note

$$\begin{aligned} A = \frac{\int_0^\infty a p_a da}{\int_0^\infty p_a da} &= \frac{\left[-a\int_a^\infty p_s ds\Big|_0^\infty + \int_0^\infty \left(\int_a^\infty p_s ds\right) da\right]}{\int_0^\infty p_a da} \\ &= \frac{\left[0 + \int_0^\infty \left(\int_a^\infty p_s ds\right) da\right]}{\int_0^\infty p_a da} \\ &= \frac{\int_0^\infty e_a^0 p_a da}{\int_0^\infty p_a da} = E. \end{aligned} \tag{5.25}$$

In the above proof, we note that expectation of life at birth (e_0^0) is represented by $e_0^0 = \int_0^\infty p_s ds$, and expected remaining life at age a is represented by $e_a^0 = \int_a^\infty p_s ds$. The hazard rate is also sometimes referred to as hazard function in survival analysis. The life table functions and survival analysis are related and frequently used in mortality analysis [33–37]. Kim and Aron, apart from their proof of $A = E$, provide a valuable analogy between life lived and life left in the context of renewal theory, which we studied in Chap. 1. Lotka's type of renewal theory was mostly mathematical, and later similar ideas were extended in probability theory that led to the renewal process. In this renewal process, a quantity known as the expected number of renewals over a time interval $[0, t]$ is established, along with concepts called *backward* and *forward recurrence* times. Let X_n represent a renewal process. For any value of $t \geq 0$, the forward recurrence time measures the time until the next renewal, while the backward recurrence time measures the time that has elapsed since the last renewal prior to t. Let the quantity $R(t)$ represent the number of renewals during $[0, t]$. The expectation of $R(t)$, i.e., $E[R(t)]$ is computed, and it forms one of the basic measures of the renewal process, called the renewal measure. The process S combined with $R(t)$ represented by $S_{R(t)}$ is used in understanding forward and backward recurrent times. A forward recurrence time process is measured through $S_{R(t)} - t$ and the backward recurrence time process is measured through $t - S_{R(t)-1}$. The age in life lived data is associated with backward recurrence time and life left data is associated with forward recurrence time of renewal process [38]. Their expression

$$\frac{E}{e_0^0} = \frac{1}{2}(1 + CV^2), \tag{5.26}$$

where CV is the coefficient of variation of the *pdf* $f(a)$ (for $F(x) = \int_0^y f(y)dy$) and $a \in X$ for a lifetime random variable X). The CV is defined as the ratio of the standard deviation of f and the expectation of the random variable. The quantity CV is a useful measure in several branches of science, including actuarial science, physics, etc. Later, several researchers used the above kind of framework in representing life lived and life left data. When $CV = 0$, we have $\frac{E}{e_0^0} = \frac{1}{2}$. Goldstein J.R. in 2009 showed the result (5.25) using an alternative approach [39]. Other contributions in the theory of life lived and life left of stationary populations include [19, 30, 31, 40–42] and of stable populations include [43].

In the following sections of the chapter, we will describe further advancements in the mathematical theory of life lived and life left data, specifically focusing on the oscillatory properties of populations, Rao's partition theorem, and alternative measures of traditional NRR.

5.5 Oscillatory Stationary Population Identity or O-SPI

In the previous sections, we saw how the SPI is derived from the life table model. In this section, we study newer extensions and advantages of the SPI, especially the oscillatory properties of Rao and Carey around stationary populations. These oscillatory properties described here are novel and bring a newer application of stationary populations [21]. Let us partition the time interval $[t_0, t_\omega)$ into two sub-collections of intervals I and J such that they are disjoint to each other. The sets I and J are defined below in (5.30) and (5.31). Let the SPI hold in each of the subintervals of the collection I, and the SPI does not hold in each of the subintervals of the collection J. The interval $[t_0, t_\omega)$ is expressed as

$$[t_0, t_\omega) = \left(\bigcup_{C_1 \in I} C_1\right) \cup \left(\bigcup_{C_2 \in J} C_2\right), \tag{5.27}$$

where C_1 and C_2 are arbitrary intervals in I and J, respectively. Instead of arbitrary collection of subintervals, later we consider a collection of subintervals within a finite time interval $[t_0, t_{k+1}] \subset [t_0, t_\omega)$. Let T be a collection of subintervals of $[t_0, t_{k+1}]$ represented by

$$T = \{[t_0, t_1), [t_1, \delta_1), [\delta_1, t_2), [t_2, \delta_2), \ldots, [\delta_k, t_{k+1}]\} \tag{5.28}$$

such that

$$[t_0, t_{k+1}] = \bigcup_{C \in T} C. \tag{5.29}$$

Further, write T as a union of I and J (i.e., $T = I \cup J$) using the elements of T in (5.28), where

$$I = [t_0, t_1) \cup \left(\bigcup_{i=1}^{k} [\delta_i, t_{i+1}) \right), \tag{5.30}$$

$$J = \bigcup_{i=1}^{k} [t_i, \delta_i). \tag{5.31}$$

Before we proceed to the formal definition of oscillatory SPI, let us define three more properties, namely, predominantly a stationary population and predominantly a nonstationary population using I, J expressed in (5.30), (5.31) and uniform amplitude of the SPI.

Definition 5.4 (Predominantly Stationary Population) A population is predominantly a stationary population if

$$I > J. \tag{5.32}$$

Definition 5.5 (Predominantly Nonstationary Population) A population is predominantly a nonstationary population if

$$I < J. \tag{5.33}$$

The SPI holds in all the subintervals of I for

$$I = \{[t_0, t_1), [\delta_1, t_2), [\delta_2, t_3), \ldots\}, \tag{5.34}$$

and the SPI does not hold in all the subintervals of J for

$$J = \{[t_1, \delta_1), [t_2, \delta_3), [t_3, \delta_4), \ldots\}. \tag{5.35}$$

Using (5.27), we define *uniform amplitude* of SPI as follows:

Definition 5.6 (Uniform Amplitude of SPI) We say the SPI has uniform amplitude if

$$[t_0, t_\omega) - \left(\bigcup_{C_1 \in I} C_1 \right) = [t_0, t_\omega) - \left(\bigcup_{C_2 \in J} C_2 \right), \tag{5.36}$$

and $[t_0, t_1) = [t_1, \delta_1), [\delta_1, t_2) = [t_2, \delta_3)$, and so on for each of the elements of I and J.

Definition 5.7 (Oscillatory-SPI) We call the SPI is oscillatory on I with uniform amplitude whenever

$$\left\{ t_1 = \frac{t_0 + \delta_1}{2}, t_i = \frac{\delta_{i-1} + \delta_i}{2} \text{ for } i = 2, 3, \dots \right\}. \tag{5.37}$$

If a population is predominantly a stationary, then during a finite small subintervals, the population could be nonstationary, and if a population is predominantly a nonstationary, then over finitely many small subintervals of time, the population could be stationary. In the next section, we will state a partition theorem of populations that proposes a criterion using the SPI to partition a large population into stationary and nonstationary components.

5.6 Rao's Partition Theorem of Populations

We saw how population that is stationary over a large time interval could be nonstationary over shorter subintervals. Building further on the SPI, we state and prove a partition theorem of population due to Arni S.R.S. Rao [15].

Theorem 5.8 (Partition Theorem, Rao [15]) *The stationary population identity (SPI) partitions a randomly selected large population into stationary and nonstationary components (if such components are non-empty).*

Proof We provide an idea of the proof. For a full technical proof refer to [15]. Let $P(t)$ be the size of a large population at time t that is distributed across all ages of a set A, where

$$A = \{0, a_1, a_2, \dots, \omega\}, \tag{5.38}$$

and

$$P(t) = \sum_{a=0}^{\omega} P_a(t). \tag{5.39}$$

Once a large population is selected and information on their ages is recorded, then each individual in the population at time t will be correspond to one of the ages in A. We compute the fraction of individuals at each age a in set A and verify from the life table if there is a fraction of the population $P(t)$ which has a years remaining to live. That is, for each a in A, we need to check if $f_1(a) = f_2(a)$, where $f_1(a)$ is the fraction of individuals in $P(t)$ who are at age a, and $f_2(a)$ is the fraction of individuals in the life table corresponding to $P(t)$ who have a years remaining to live.

For the set of elements in A for which $f_1(a) = f_2(a)$ is satisfied, we call the corresponding population as a stationary component. For the rest of the elements in A for which $f_1(a) \neq f_2(a)$, we call the corresponding population as a nonstationary component of $P(t)$. Hence, we could write

$$P(t) = M(t) + N(t), \tag{5.40}$$

where $M(t)$ is the stationary component and $N(t)$ is the nonstationary component of $P(t)$ at time t. □

Once $P(t)$ is partitioned, the set A can be written as $A = S_1 \cup S_2$, where S_1 consists of set of all ages in A that satisfies $f_1(a) = f_2(a)$, and S_2 consists of set of all ages in A that satisfies $f_1(b) \neq f_2(b)$. See Fig. 5.4. We can summarize this as

$$S_1 : \{a : a \in A \text{ and } f_1(a) = f_2(a)\}, \tag{5.41}$$

$$S_2 : \{b : b \in A \text{ and } f_1(b) \neq f_2(b)\}. \tag{5.42}$$

We note here that the definitions of functions f_1 and f_2 in Rao's partition theorem differ from those used in the proof of the Rao–Carey fundamental theorem described in Sect. 5.3.

Example 5.9 Let $A = \{0, 1, 2, \ldots, 6\}$ be the set of all ages of a population. Let B be the set of fractions of individuals in the population at ages listed in A, and C be the set of fractions of individuals who have a years remaining to live for a in A. Let f and g represent fractions in B and C, respectively. Suppose

$$B = \{f(0) = 0.3, f(1) = 0.25, f(2) = 0.15, f(3) = 0.12,$$
$$f(4) = 0.1, f(5) = 0.06, f(6) = 0.02\},$$

$$C = \{g(0) = 0.37, g(1) = 0.1, g(2) = 0.15, g(3) = 0.18,$$
$$g(4) = 0.1, g(5) = 0.07, g(6) = 0.03\}.$$

The graph based on the three sets A, B, and C is shown in Fig. 5.5.

5.6.1 *Advantages of Partition Theorem*

The ideas of the partition theorem presented in this section are emerging in the demographic literature recently, which we will see later in the chapter. Such a partition theorem may be a way of concluding whether a given country or a region attained a population replacement level, which is argued as a better method than

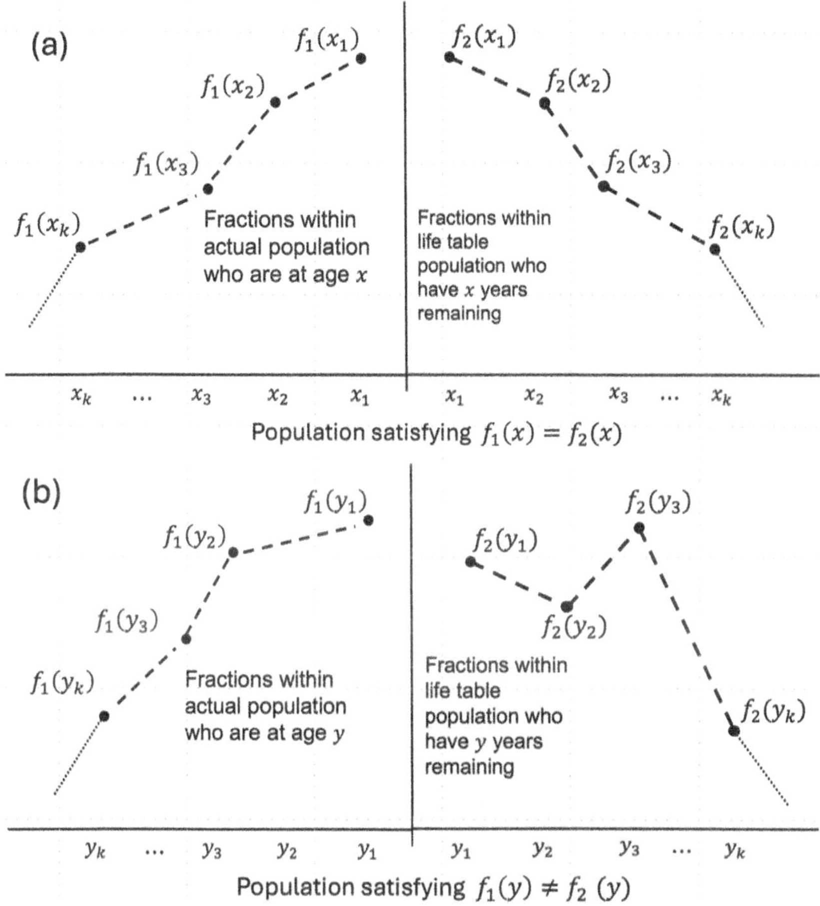

Fig. 5.4 Stationary and nonstationary components of a large population. (**a**) The set of all ages in A that satisfies $f_1(x) = f_2(x)$ and (**b**) the set of all ages in A that satisfies $f_1(y) \neq f_2(y)$

computing the replacement level through the traditional NRR formula in (5.43). Refer to [13, 14]. Next, we will discuss these new and alternative formulae for computing NRR. Partition theorem-based conclusions on whether a country attained stationary status are more time-sensitive than traditional NRR-based judgment of population replacement levels. We end the section with two examples of the interpretation of the partition theorem discussed here; see Example 5.10. In Sect. 5.8, we will study application of the partition theorem to other situations and for continuous time.

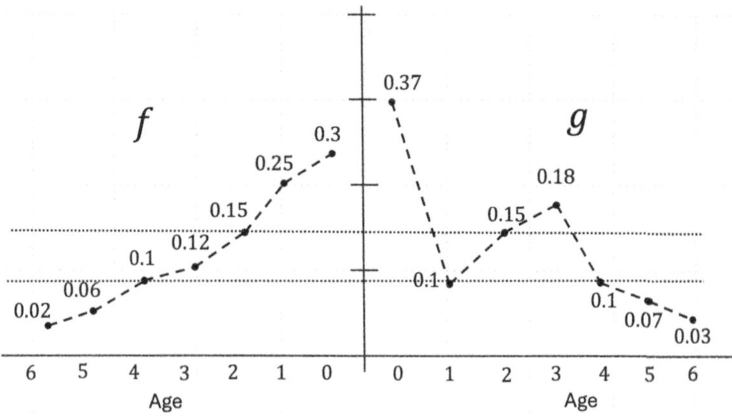

Fig. 5.5 Fraction of individuals at each age in the population versus fraction of individuals who have a remaining years for a in A in Example 5.9

Example 5.10 For a country's population at time t or for a region in a country at time t, let

$$S_1 = \{5, 6, 12, 15, 17, 43, 55, 84, 89, 103\}$$

and $\omega = 105$. Note that $S_1 : \{x : x \in A \text{ and } f_1(x) = f_2(x)\}$ described in (5.41). Assume life tables are available for this population at time t. Find the stationary and nonstationary components of the population based on S_1, if the total population is assumed to be partitioned by the partition theorem above.

Solution We have seen that S_1 is the set of all ages for which the fraction of individuals in the population who are at age x is equal to the fraction of individuals in the corresponding life table population who are at age x, i.e., $f_1(x) = f_2(x)$. These are explained above in Sect. 5.6. Specifically, for the ages in S_1, the equality $\frac{P_5}{P(t)} = \frac{L_5}{L(t)}, \frac{P_6}{P(t)} = \frac{L_6}{L(t)}, \ldots$ holds. See Fig. 5.6. For all other ages for which this equality is not true, the set S_2 holds. Hence, S_2 can be determined by $S_2 = A - S_1$. The age set A was defined in (5.6). We know from the description above, $P(t) = M(t) + N(t)$, where $M(t)$ is the stationary component and $N(t)$ is the nonstationary component of of the total population $P(t)$. Since for the people who are in the ages listed in S_1, i.e., $P_5, P_6, P_{12}, \ldots, P_{103}$ the equality $f_1(x) = f_2(x)$ holds, where P_a is the population who are at age a. For all other ages, $f_1(y) \neq f_2(y)$ holds. Therefore, the stationary component is represented by

$$M(t) = P_5 + P_6 + \ldots + P_{103},$$

and the nonstationary component is represented by

$$N(t) = P(t) - M(t) = P(t) - \{P_5 + P_6 + \ldots + P_{103}\}.$$

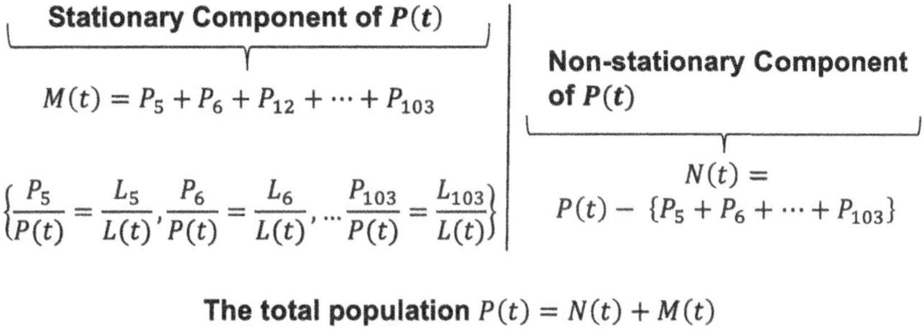

Stationary Component of $P(t)$

$$M(t) = P_5 + P_6 + P_{12} + \cdots + P_{103}$$

$$\left\{\frac{P_5}{P(t)} = \frac{L_5}{L(t)}, \frac{P_6}{P(t)} = \frac{L_6}{L(t)}, \ldots \frac{P_{103}}{P(t)} = \frac{L_{103}}{L(t)}\right\}$$

Non-stationary Component of $P(t)$

$$N(t) = P(t) - \{P_5 + P_6 + \cdots + P_{103}\}$$

The total population $P(t) = N(t) + M(t)$

Fig. 5.6 Numerical demonstration of the application of Rao's partition theorem. See Example 5.10. The total population $P(t)$ is partitioned into a stationary component ($M(t)$) and nonstationary component ($N(t)$), where $P(t) = M(t) + N(t)$. The life table population at time t is $L(t)$ and $L_0, L_1, \ldots, L_\omega$ are the single-age life table populations of $L(t)$

Example 5.11 Let us consider the population in Example 5.10, and $P(t) = 102{,}527$. Suppose

$$f_1(5) = 0.01, f_1(6) = 0.01, f_1(12) = 0.02, f_1(15) = 0.007, f_1(17) = 0.008,$$
$$f_1(43) = 0.008, f_1(55) = 0.006, f_1(84) = 0.006, f_1(89)0.005, f_1(103) = 0.001.$$

Find the values of stationary ($M(t)$) and non-stationary ($N(t)$) components of $P(t)$.

Solution The stationary component of the total population is obtained by summing the fractions $f_1(5), f_1(6), \ldots, f_1(103)$ and multiplying the sum with the given total population size. Remember these fractions given are for the ages in the set S_1 in Example 5.10. The set satisfies $f_1(a) = f_2(a)$ of the partition theorem in this section. That is,

$$M(t) = [f_1(5) + f_1(6) + \ldots + f_1(103)] \times P(t) = 0.081 \times 102{,}527 \approx 8305,$$
$$N(t) = P(t) - M(t) = 102{,}527 - 8305 = 94{,}222.$$

5.7 Rao's Alternative Formula to Compute NRR

We first discuss the limitations of the traditional NRR formula introduced in Chap. 1 (see 5.43) and then suggest alternative formulae. Given that the NRR is a popular demographic measure and measuring population replacement levels is an important exercise carried out by several agencies in the world, it is important to know that the NRR value computed for a country based on the data in a particular year does not reflect the population replacement level of that country in that year.

Using (1.10) (in Chap. 1) the discrete equivalent of R_0 is expressed as

$$R_0 = \sum_{a=\alpha}^{\beta} l(a)m(a), \tag{5.43}$$

where $l(a) = L_a^t/l_0$ is obtained from a life table constructed for the year t. The quantity L_a^t is the life table population of women at age a obtained from the single-age life table constructed for the year t with a radix of l_0=100,000. More on life tables are in Chap. 3. The maternity function is computed from the formula $m(a) = B_a^t/W_a^t$, where B_a^t is the total number of female children born to the women of age a for the year t, and W_a^t is the effective number of women of age a for the year t.

New formulae were based on a technical note in *PAA Affairs* [13] and on the presentation at the *IPC 2021 Conference* (online), Hyderabad, India [14]. The traditional NRR formula can be written as follows:

$$R_0 = \sum_{a=15}^{49} \frac{B_a^t}{W_a^t} \frac{L_a^t}{l_0}, \tag{5.44}$$

where

W_a^t : the number of women who are at age a in the year t

B_a^t : the total number of femile children born to W_a^t

L_a^t : the female life table population who are at age a in the year t

l_0 : radix or a cohort of a life table

The quantities B_a^t and W_a^t are often obtained from demographic surveys conducted in the year t and L_a^t and l_0 are obtained from a female life table computed for the year t. When $R_0 = 1$, we conventionally say that the population attained the replacement level, i.e., on an average a female in her lifetime replaced a girl child. We also saw the arguments earlier that if $R_0 > 1$, the population grows, and if $R_0 < 1$, then the population decays. We discussed R_0 earlier with respect to the stable population theory of Lotka. In general, if the population is not stable, the NRR value of 1 computed through the formula in (5.44) need not indicate anything about the actual population replacement level in the year it was computed. The NRR value in (5.44) only indicates the futuristic replacement rate because the quantity L_a^t/l_0 that was used indicates the chance of survival of a newborn girl in the year t to survive until age a based on the age-specific mortality rates in all the ages in the female population that were collected in the year t. Hence, even if $R_0 = 1$ is obtained in a year that does not guarantee that the population is already at the replacement level in the year t or it will surely attain the replacement level sometime after the year t. This is a serious limitation in the interpretation of the traditional method of computing population replacement levels using (5.43). The age-specific mortality rates of a population may not be stable and they might change in the future after they were computed in the year t. This limitation contributes to uncertainty in estimating the population replacement level. At the same time, it is impossible to predict accurately future age-specific mortality rates of a population and compute future life tables. At least theoretically, one can provide alternative formulae that can compute NRR values by overcoming these limitations discussed.

We now introduce a new formula to compute the retrospective replacement level of fertility and denote it by Q_0. Refer to [13] for more details. As defined earlier, let α and β be the lower and upper ages of reproduction of women, respectively. The formula for Q_0 is

$$Q_0 = \sum_{a=\alpha}^{\beta} \frac{B_a^t}{W_a^t} \frac{F_a^{t+a}}{B_0^t}, \tag{5.45}$$

where the quantities B_x^t and W_a^t have the same meaning as in (5.44), B_0^t is the total new girl babies born in the year t, and F_a^{t+a} is the effective person-years lived by the survivors of B_0^t who are currently of age a in the year $t + a$. The female population in the year $t + a$ was born in the year t. For the effective person-years in a closed population, if $Q_0 = 1$ in the year $t + a$, we can conclude that the population of the year t reached replacement-level fertility in the year $t + a$ for $a \in [\alpha, \beta]$. For any year $t + a$, the values of F_a^{t+a} can be computed without any prediction tools. See Table 5.1 for an example computation of Q_0. The following are two key advantages of Q_0 in (5.45) over R_0 expressed in (5.44): (i) There is no need for construction of life tables for the year t or for any year $t + a$ for $a \in [\alpha, \beta]$, and (ii) unlike some uncertainty associated with R_0 due to L_a^t/l_0 and assumption of constant age-specific mortality rates, the quantity F_a^{t+a}/B_0^t is not influenced by uncertainty (if obtained from an accurate data).

Example 5.12 See Table 5.1. There are 147 births delivered by 10,000 women in the year t. The distribution of births by age distribution of women is given in the table. Let the number of deaths among 147 until the age 15, during 15–23, 23–29, 29–36, and 36-46, be 7, 7, 8, 7, and 15, respectively. Assuming uniform distribution

Table 5.1 Computation of Q_0 (with hypothetical data). See Example 5.12 for a related description

(1) Age	(2) B_a^t	(3) W_a^t	(4) F_a^{t+a}	(5) B_a^t/W_a^t	(6) F_a^{t+1}/B_0^t	(7) (5)×(6)
15	43	2800	2150	0.015	14.63	0.225
23	37	2344	3242	0.016	22.05	0.348
29	39	2010	4016	0.019	27.34	0.530
36	21	1620	4866	0.013	33.10	0.429
46	7	1226	5971	0.006	40.62	0.232
	$B_0^t = 147$					$Q_0 = 1.764$

of deaths, the person-years column F_a^{t+a} is computed. For example, $F_{15}^{t+15} = 147 \times 15 - 7 \times (15/2) \approx 2150$, $F_a^{t+23} = 2150 + 140 \times 8 - 7 \times (23-15)/2 \approx 3242$, $F_a^{t+29} = 3242 + 133 \times 6 - 8 \times (29-23)/2 = 4016$, $F_a^{t+36} = 4016 + 125 \times 7 - 7 \times (36-29)/2 = 4866$, $F_a^{t+46} = 4866 + 118 \times 10 - 15 \times (46-36)/2 = 5971$. The value of $Q_0 = 1.764$ indicates the female population of the year t is replaced by ≈ 1.8 females in the year $t + 46$. One can compute F_a^{t+a} using other kinds of assumptions on death distributions based on the evidence collected.

5.8 Partition Theorem on a Continuous Age Interval

Let us consider a continuous age interval $[0, \omega]$ and define two functions f_3 and f_4 as follows:

$$f_3 : [0, \omega] \to \frac{L(a,t)}{\int_0^\infty L(a,t)da}, \tag{5.46}$$

$$f_4 : [0, \omega] \to \frac{P(a,t)}{\int_0^\infty P(a,t)da}, \tag{5.47}$$

where $L(a,t)$ is the life table population at age a and time t, and $P(a,t)$ is the actual (census) population at age a and time t. Let us partition $[0, \omega]$ into three sets U, V, and W such that

$$U = \left\{ a : \left(\frac{L(a,t)}{\int_0^\infty L(a,t)da} > \frac{P(a,t)}{\int_0^\infty P(a,t)da} \right) \forall a \in [0, \omega] \right\},$$

$$V = \left\{ a : \left(\frac{L(a,t)}{\int_0^\infty L(a,t)da} < \frac{P(a,t)}{\int_0^\infty P(a,t)da} \right) \forall a \in [0, \omega] \right\},$$

and

$$W = \left\{ a : \left(\frac{L(a,t)}{\int_0^\infty L(a,t)da} = \frac{P(a,t)}{\int_0^\infty P(a,t)da} \right) \forall a \in [0, \omega] \right\}. \tag{5.48}$$

We will state and prove a theorem that establishes a relationship between f_3 and f_4 using U, V, and W. This relationship aids in partitioning populations and employs principles of mathematical analysis.

Theorem 5.13 (Rao and Carey, 2023 [23])

$$\int_0^{\infty}\left(\frac{L(a,t)}{\int_0^{\infty}L(a,t)da}-\frac{P(a,t)}{\int_0^{\infty}P(a,t)da}\right)da=0, \tag{5.49}$$

if, and only if,

$$\int_{a\in U}\left(\frac{L(a,t)}{\int_0^{\infty}L(a,t)da}-\frac{P(a,t)}{\int_0^{\infty}P(a,t)da}\right)da=$$
$$-\int_{a\in V}\left(\frac{L(a,t)}{\int_0^{\infty}L(a,t)da}-\frac{P(a,t)}{\int_0^{\infty}P(a,t)da}\right)da. \tag{5.50}$$

Proof We will first prove forward part of the statement. Let us assume (5.49) is true. We will prove (5.50). The two elements $\{0\}$ and $\{\omega\}$ could be in any of the sets U, V, and W described in (5.48). We list below all possibilities of locations of $\{0\}$ and $\{\omega\}$:

(i) $0 \in U$ and $\omega \in U$,

(ii) $0 \in U$ and $\omega \in V$,

(iii) $0 \in U$ and $\omega \in W$,

(iv) $0 \in V$ and $\omega \in V$,

(v) $0 \in V$ and $\omega \in U$,

(vi) $0 \in V$ and $\omega \in W$,

(vii) $0 \in W$ and $\omega \in W$,

(viii) $0 \in W$ and $\omega \in U$,

(ix) $0 \in W$ and $\omega \in V$.

Consider case (i). Since $0 \in U$, we have

$$\frac{L(0,t)}{\int_0^{\infty}L(a,t)da}>\frac{P(0,t)}{\int_0^{\infty}P(a,t)da},$$

which implies $f_4(0) < f_3(0)$. The function f_4 touches f_3 at some age, say a_1 from the bottom of f_3. That is, $f_3(a_1) = f_4(a_1)$. For some a immediately after a_1 ($a > a_1$), we will have three situations: $f_4 > f_3$, or $f_4 = f_3$, or $f_4 < f_3$.
Suppose $f_4 > f_3$: Given that $\omega \in U$, the function f_4 has to reach f_3 from the bottom of f_3 so even if $f_4 > f_3$ for some age $a > a_1$, the function f_4 has to start declining from some age, say a_2, and touch f_3 at age, say a_3, from the above of f_3

and reach f_3 at ω such that

$$\int_0^{a_1}\left(\frac{L(a,t)}{\int_0^\infty L(a,t)da}-\frac{P(a,t)}{\int_0^\infty P(a,t)da}\right)da+$$
$$\int_{a_3}^{\omega}\left(\frac{L(a,t)}{\int_0^\infty L(a,t)da}-\frac{P(a,t)}{\int_0^\infty P(a,t)da}\right)da=$$
$$-\int_{a_1}^{a_2}\left(\frac{L(a,t)}{\int_0^\infty L(a,t)da}-\frac{P(a,t)}{\int_0^\infty P(a,t)da}\right)da-$$
$$\int_{a_2}^{a_3}\left(\frac{L(a,t)}{\int_0^\infty L(a,t)da}-\frac{P(a,t)}{\int_0^\infty P(a,t)da}\right)da. \tag{5.51}$$

Equality (5.51) holds because of hypothesis (5.49).
Suppose $f_4 = f_3$: Suppose $f_4 = f_3$ for some time after $a > a_1$, and f_4 becomes less than f_3 at some age, say a_2. Then f_4 cannot reach f_3 at ω from the bottom of f_3 without having a situation of $f_4 > f_3$ for an age interval because of hypothesis (5.49). We can bring an argument such that (5.51) kind of equation holds. Suppose $f_4 = f_3$ for some time after $a > a_1$, and f_4 becomes higher than f_3 at some age, say a_2; then we will repeat the arguments that will arrive a kind of (5.51) with some disjoint age intervals.
Suppose $f_4 < f_3$: In this situation, f_4 cannot reach f_3 at ω from the bottom without again crossing the function f_3 at some age because of the reason explained previously. Hence, we will have a situation for some ages such that (5.51) kind of equation holds.

Suppose f_4 oscillates around f_3 as shown in Fig. 5.7, given hypothesis (5.49), we will have

$$\left\{\int_0^{a_1}+\int_{a_3}^{a_4}+\int_{a_4}^{a_5}+\int_{a_7}^{a_8}+\int_{a_8}^{a_9}+\int_{a_{11}}^{a_{12}}+\int_{a_{12}}^{a_\omega}\right\}\times$$
$$\left(\frac{L(a,t)}{\int_0^\infty L(a,t)da}-\frac{P(a,t)}{\int_0^\infty P(a,t)da}\right)da=$$
$$-\left\{\int_{a_1}^{a_2}+\int_{a_2}^{a_3}+\int_{a_5}^{a_6}+\int_{a_6}^{a_7}+\int_{a_9}^{a_{10}}+\int_{a_{10}}^{a_{11}}\right\}\times$$
$$\left(\frac{L(a,t)}{\int_0^\infty L(a,t)da}-\frac{P(a,t)}{\int_0^\infty P(a,t)da}\right)da. \tag{5.52}$$

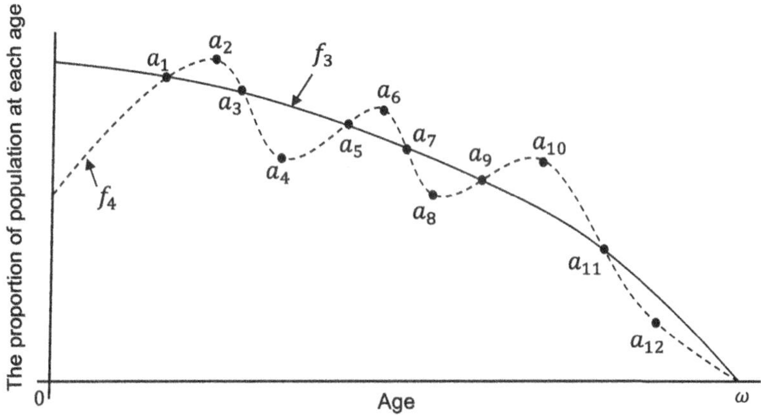

Fig. 5.7 Relationship between the proportions of life table populations (f_3) and actual populations (f_4)

This leads to

$$\sum_{i=0}^{13} \int_{a_i}^{a_{i+1}} \left(\frac{L(a,t)}{\int_0^\infty L(a,t)da} - \frac{P(a,t)}{\int_0^\infty P(a,t)da} \right) da = 0, \tag{5.53}$$

where $a_0 = 0$, and $a_{13} = \omega$.

For all other combinations of the relationship between f_3 and f_4, Eq. (5.52) will be reached. Hence, (5.50) is true. Similarly, for each of the cases (ii)–(ix), the above kind of arguments can be constructed to prove the forward statement of the theorem.

For the "only if" part of the proof, let us assume (5.50) holds. We now will prove (5.49). Since (5.50) holds, we will have

$$\int_{a\in U} \left(\frac{L(a,t)}{\int_0^\infty L(a,t)da} - \frac{P(a,t)}{\int_0^\infty P(a,t)da} \right) da + \int_{a\in V} \left(\frac{L(a,t)}{\int_0^\infty L(a,t)da} - \frac{P(a,t)}{\int_0^\infty P(a,t)da} \right) da = 0. \tag{5.54}$$

For all $a \in W$, we have $f_3(a) = f_4(a)$, so

$$\int_{a\in W} \left(\frac{L(a,t)}{\int_0^\infty L(a,t)da} - \frac{P(a,t)}{\int_0^\infty P(a,t)da} \right) da = 0. \tag{5.55}$$

From (5.54) and (5.55), we arrive at (5.49). □

5.9 Discrete Population Capture Models

This section is developed based on geometric shape formations and associated mathematical analysis while constructing capture cohorts of individuals in demography, especially in demographic field experiments. Refer to [17]. The ideas developed were new to the mathematical demography literature of stationary populations. The ideas of Rao in [37] were presented here in a more accessible style. Geometrical intuition has played an interesting and useful role in population analysis as we saw in Sects. 5.1 and 5.2. The intuitions with which the recent results built in the demography of captive cohort formations help us better understand the speed of cohort formation in field experiments, the size of a sub-cohort within a total cohort, the design of field demographic experiments, etc.

In a stationary population, let us randomly collect individuals to form a capture cohort that is represented by

$$c(t_0) = \int_0^\infty n(a, t_0)da, \tag{5.56}$$

where $n(a, t_0)$ is the number of captured individuals at age a and at time t_0. The cohort $c(t_0)$ is formed out of individuals of all ages in that population. After formation of initial cohort at t_0, we define two more cohorts $c(s)$ and $c(\infty)$, representing capture cohorts formed during time intervals (t_0, s) and (t_0, ∞), respectively. These two cohorts are expressed as

$$c(s) = \int_0^s \int_0^\infty n(a, t_b)dadb, \tag{5.57}$$

$$c(\infty) = \int_0^\infty \int_0^\infty n(a, t_b)dadb. \tag{5.58}$$

Here $c(\infty)$ can be treated as

$$c(\infty) = \lim_{s\to\infty} c(s) \tag{5.59}$$

indicating a continuous formation of the cohort beyond the time s. The size of the cohort $c(\infty)$ will be larger than the cohort $c(s)$ (in the absence of any mortality during (s, ∞)). It was also assumed that each individual within these cohorts has an equal probability to get captured. This will lead us to McKendric von Foerster type of partial differential equations (PDEs) that govern the dynamics of the cohorts $c(s)$ and $c(\infty)$ and are given by

$$\begin{aligned} \frac{\partial c(s)}{\partial a} &= v_1 \frac{\partial c(s)}{\partial b}, \\ \frac{\partial c(s)}{\partial b} &= - v_2 \frac{\partial c(s)}{\partial a}, \end{aligned} \tag{5.60}$$

where the parameters can be modeled by

$$v_1 = v_2 = \int_{a\in\mathbb{R}^+} a\left(\frac{c(s)}{\int_0^\infty n(a,t_0)da}\right)da.$$

In [17], more properties of population dynamics through such models were studied using Lebesgue measures. See Appendix F for a definition of Lebesgue measure. We will discuss here some essential features and model-building points. Elsewhere in the book (see Chap. 2), we studied McKendrik-von-Foerster-type basic population model formation. Using the ideas of $c(s)$ described above, a collection of discrete cohorts C is formed, where

$$C = \{c(s_0, s_1), c(s_1, s_2), \ldots, c(s_{n-1}, s_n)\}. \tag{5.61}$$

In (5.61),

$$c(s_{n-1}, s_n) = \int_{s_{n-1}}^{s_n}\int_0^\infty n(a, t_b)dadb \tag{5.62}$$

for $i = 1, 2, \ldots, n$ formed within $(t_{s_{n-1}}, t_{s_n})$. The interval (t_0, t_n) is partitioned into

$$\left\{[t_0, t_1], (t_1, t_2], \ldots, (t_{s_{n-1}}, t_{s_n}]\right\} \tag{5.63}$$

subintervals. The capture cohorts within the subintervals (5.63) need not be of the same size. The above collection of sets C could be formed either by fixing the subinterval duration (Length of the subinterval) of capturing or based on the size of capture cohort predetermined. There are three following types of discrete cohorts formed:

(i) The size of $n_0(s_{i-1}, s_i)$ is random and the durations of lengths in (5.63) are constant.
(ii) The size of $n_0(s_{i-1}, s_i)$ is predetermined and the durations of lengths in (5.63) are constant.
(iii) The size of $n_0(s_{i-1}, s_i)$ is predetermined and the total cohort size $c(s)$ is also predetermined.

Theorem 5.14 (Based on a Proposition in Rao [17]) *In an infinitely large population, let $S_1, S_2, \ldots, S_n$ be n geographic regions in which demographic experiments are conducted. Let $c(s, S_1), c(s, S_2), \ldots, c(s, S_n)$ be the cohort sizes formed in n regions. Suppose $n[S_k, (s_{i-1}, s_i)]$ exists for $k = 1, 2, \ldots, k$ and $i = 1, 2, \ldots, n$ under type (ii) discrete cohort such that*

$$n[S_1, (s_{i-1}, s_i)] = n[S_2, (s_{i-1}, s_i)] = \cdots = n[S_m, (s_{i-1}, s_i)],$$

and $n\left[A_k, (s_{i'-1}, s_{i'})\right]$ *exists for* $k = 1, 2, \ldots, k$ *and* $i' = 1, 2, \ldots, n'$ *under type (iii) discrete cohort such that*

$$n\left[A_1, (s_{i'-1}, s_{i'})\right] = n\left[A_2, (s_{i'-1}, s_{i'})\right] = \cdots = n\left[A_m, (s_{i'-1}, s_{i'})\right],$$

and $n = n'$.

Then

$$\sum_{k=1}^{m} c(\infty, A_k) \, [\textit{under type (ii) discrete cohort}] \leq$$

$$\sum_{k=1}^{m} c(\infty, A_k) \, [\textit{under type (iii) discrete cohort}]. \tag{5.64}$$

Proof We provide here an idea of the proof from [17]. Under the hypothesis, let us define two events E_1 and E_2 as follows:

E_1 : $\{n\,[S_k, (s_{i-1}, s_i)]$ exists under type (ii) cohorts for all i and k, and the quantity $n\left[S_k, (s_{i'-1}, s_{i'})\right]$ exists under type (iii) cohorts for all i' and k and $n = n'$, E_2 : Some of $n\,[S_k, (s_{i-1}, s_i)]$ do not exist under type (ii) but the quantity $n\left[S_k, (s_{i'-1}, s_{i'})\right]$ exists under type (iii) for i' and k, and $n = n'\}$. The size of the cohort $c(s, S_i)$ is defined as

$$c(s, S_i) = \int_0^\infty n(a, t_b, S_i) dadb, \tag{5.65}$$

where $n(a, t_b, S_i)$ is the number of individuals in age a at time t_b in the region S_i. When the event E_1 occurs, we have

$$\begin{aligned} c(s, S_1) &= c(s, S_2) = \ldots = c(s, S_m) \\ &\Rightarrow \lim_{s\to\infty} c(s, S_1) = \lim_{s\to\infty} c(s, S_2) = \ldots = \lim_{s\to\infty} c(s, S_m). \\ &\Rightarrow c(\infty, S_1) = c(\infty, S_2) = \ldots = c(\infty, S_m). \end{aligned}$$

This implies

$$\sum_{k=1}^{m} c(\infty, S_k) \left[\text{under type (ii) discrete cohort}\right] =$$

$$\sum_{k=1}^{m} c(\infty, S_k) \left[\text{under type (iii) discrete cohort}\right]. \tag{5.66}$$

When E_2 occurs, the size of the cohorts formed under type (ii) would be less than the cohorts formed under type (iii). Hence,

$$\sum_{k=1}^{m} c(\infty, S_k) \left[\text{under type (ii) discrete cohort}\right] <$$

$$\sum_{k=1}^{m} c(\infty, S_k) \left[\text{under type (iii) discrete cohort}\right]. \tag{5.67}$$

Using (5.66) and (5.67), we arrive at (5.64). □

Chapter 5 describes several new discoveries and theories of stationary populations since the Euler–Lotka stationary population models. The definition of the stationary population concerning a life table was not changed in these discoveries. Still, newer implications of stationary population identity (derived from a life table context of a stationary population) were brought into the literature on the mathematics of stationary population. These discoveries include proof of the stationary population identity using algebraic, combinatorics arguments (Rao–Carey fundamental theorem) and oscillatory properties of populations between nonstationary and stationary and partition theorem of populations (Rao's partition theorem). We saw in the chapter that the partition theorem can instantly divide the total population into stationary and nonstationary component. A major advancement in applications of stationary populations of mathematical demography is due to partition theorems.

The chapter also provides an alternative and somewhat more elegant formula to compute NRR (Rao, 2021 [13]) so that the population replacement level in a country or a region can be understood promptly. The standard NRR formula using a discrete form that was derived from Lotka's continuous form is mathematically correct but has limitations in predicting the population replacement level in the year for which it was computed. These discussions can be seen in Sect. 5.7.

5.10 Exercises

Exercise 5.15 Let x and y be *units of time* described in equality (5.2), but in continuous form, and let $n(x)$ and $m(y)$ be their corresponding population frequencies at x and y, respectively. Let

$$A_1 = \int_0^{\omega} n(x)dx,$$

$$B_1 = \int_0^{\omega} xn(x)dx,$$

$$C_1 = {}^{B_1}/_{A_1},$$

$$A_2 = \int_0^{\omega} m(y)dy,$$

$$B_2 = \int_0^{\omega} ym(y)dy,$$

and

$$C_2 = {}^{B_2}/_{A_2}.$$

From the results discussed in this chapter, prove the following two statements: In a stationary population, (i) $A_1 = A_2$ and $B_1 = B_2 \implies C_1 = C_2$ and (ii) $A_1 = A_2$ and $C_1 = C_2 \implies B_1 = B_2$.

Exercise 5.16 Let us consider a country with a total population of $P(t) = 97{,}846$. Suppose the total population of the country can be partitioned according to Rao's partition theorem of populations into stationary and nonstationary components. Let A be the set of all single-year ages in the population represented by $A = \{0, 1, 2, \ldots, 105\}$. Let us consider two fractions $f_1(a)$ and $f_2(a)$, where $f_1(a)$ is the fraction of individuals in $P(t)$, and $f_2(a)$ is the fraction of individuals in the corresponding life table for a years remaining to live for $a \in A$. Suppose $A = S_1 \cup S_2$, where the set S_1 consists of all ages in A that satisfies $f_1(a) = f_2(a)$ in $P(t)$, and the set S_2 consists of all ages in A that satisfies $f_1(b) \neq f_2(b)$. That is,

$$S_1 : \{a : a \in A \text{ and } f_1(a) = f_2(a)\},$$
$$S_2 : \{b : b \in A \text{ and } f_1(b) \neq f_2(b)\}.$$

Given that $S_1 = \{5, 13, 18, 31, 49, 80\}$ and $f_1(5) = 0.02$, $f_1(13) = 0.01$, $f_1(18) = 0.007$, $f_1(31) = 0.007$, $f_1(49) = 0.006$, $f_1(80) = 0.006$. Find the stationary and the nonstationary components of the total population.

Exercise 5.17 For a stationary population, let the average age be equal to the average age of the expected remaining life. Let X be a random variable representing the duration of a lifetime. Find an expression for the coefficient of variation for the probability density function $f(x)$ for $x \in X$.

Exercise 5.18 Using Table 5.2, compute the value of Q_0. Here Q_0 is an alternative formula for the traditional R_0 value described in Sect. 5.7. The distribution of deaths among 224 births in the respective ages in Table 5.2 are 10, 10, 13, 9, and 23. Assume uniform distribution of deaths within a year.

Exercise 5.19 Obtain an expression for the expectation of life at birth for a year t, given that there were $\int_0^{\infty} c(a)da$ births in the year t. Assume that $\int_{na}^{(n+1)a} d(s)ds$ number of total deaths occur uniformly for $n = 0, 1, 2, \ldots \infty Hint.$ Consider the intervals $\{(0, a), (a, 2a), \ldots.$ and obtain an expression for person-years lived during each of these intervals.

Table 5.2 Births and women distribution

(1) Age	(2) B_a^t	(3) W_a^t
15	65	4200
24	57	3516
27	59	3015
35	32	2430
44	11	1839
	$B_0^t = 224$	$\sum_a W_a^t = 15{,}000$

Exercise 5.20 Define S_1 and S_2 as per the Exercise 5.16 and let

$$S_1 = \{1, 9, 13, 21, 28, 41, 59, 71, 89, 101\}$$

and $f_1(1) = 0.02$, $f_1(9) = 0.01$, $f_1(13) = 0.02 f_1(21) = 0.009$, $f_1(28) = 0.009$, $f_1(41) = 0.007$, $f_1(59) = 0.007$, $f_1(71) = 0.006$, $f_1(89) = 0.003$. Assume that the total population can be partitioned per Rao's partition theorem. Find the stationary and the nonstationary components of the total population.

Exercise 5.21 Let $A = \{0, 1, 2, \ldots, 9\}$ be the set of all ages of a population. Let the sets B and C represent the fractions (f) of individuals in the population at ages listed in A and fractions (g) of individuals who have a years remaining to live for a in A, respectively. Suppose

$$B = \{f(0) = 0.18, f(1) = 0.15, f(2) = 0.11, f(3) = 0.12, f(4) = 0.1,$$
$$f(5) = 0.06, f(6) = 0.02, f(7) = 0.12, f(8) = 0.1, f(9) = 0.04\},$$

$$C = \{g(0) = 0.23, g(1) = 0.1, g(2) = 0.11, g(3) = 0.18, g(4) = 0.1,$$
$$g(5) = 0.07, g(6) = 0.02, g(7) = 0.15, g(8) = 0.03, g(9) = 0.01\}.$$

Draw a graph similar to Fig. 5.4 based on the three sets A, B, and C, and show the stationary and nonstationary components' age groups in the population.

References

1. Smith, D.P., Keyfitz, N.: Mathematical Demography. Selected papers. Second, revised edition. Edited and with commentaries by Kenneth W. Wachter and Hervé Le Bras. Demographic Research Monographs, xxiv+335 pp. Springer, Heidelberg, (2013). ISBN: 978-3-642-35857-9; 978-3-642-35858-6
2. Halley, E.: An estimate of the degrees of the mortality of mankind, drawn from curious tables of the births and funerals at the city of Breslaw; with an attempt to ascertain the price of annuities upon lives. Phil. Trans. Roy. Soc. Lond. **17**, 596–610 (1693)

3. Euler, L.: Recherches gfinfirales sur la mortalitfi et la multiplication du genre humain. Hist. Acad. R. Sci. B.-Lett. Berl **16**, 144–164 (1760/1767). eulerarchive.maa.org
4. Euler, L.: Sur la multiplication du genre humain. In: Leonhardi Euleri Opera omnia, Ser. I, vol. 7, pp. 545–552. Teubner, Leipzig (1923)
5. Euler, L.: Introductio in analysin infinitorum, Tomus primus. Bousquet, Lausanne (1748). Leonhardi Euleri Opera omnia, Ser. I, vol. 8. Teubner, Leipzig (1922)
6. Krantz, S.G., Rao, A.S.R.S.: Ancient Indian mathematics needs an honorific place in modern mathematics celebration. Eur. Math. Soc. Mag. **130**, 36–39 (2023)
7. Rao, A.S.R.S., Krantz, S.G.: Dynamical systems: From classical mechanics and astronomy to modern methods. J. Ind. Inst. Sci. **101**(3), 419–429 (2021)
8. Shylaja, B.S., Shubha, B.S.: Retrograde motion as described in Brahmatulya-udāharaṇam of Viśvanātha. Ind. J. Hist. Sci. **55**(1), 40–48 (2020)
9. Kotoulas, T., Morais, M.H.M., Voyatzis, G.: The phase space structure of retrograde mean motion resonances with Neptune: the 4/5, 7/9, 5/8 and 8/13 cases. Celestial Mech. Dynam. Astronom. **134**, **6**, Paper No. 52, 25 pp. (2022)
10. Chabás, J., Goldstein, B.R.: Ibn al-Kammād's Muqtabis zij and the astronomical tradition of Indian origin in the Iberian Peninsula. Archive History Exact Sci. **69**, 577–650 (2015)
11. Aaboe, A.: Episodes from the Early History of Astronomy. Springer, New York and Berlin (2001)
12. Rao, A.S.R.S.: Population stability and momentum. Notic. Am. Math. Soc. **61**(9), 1062–1065 (2014). (https://www.ams.org/notices/201409/rnoti-p1062.pdf)
13. Rao, A.S.R.S.: Is NRR Time-Sensitive in Measuring Population Replacement Level. PAA Affairs. PAA: Population Association of America (2021). https://www.populationassociation.org/blogs/emily-merchant1/2021/01/26/is-nrr-time-sensitive
14. Rao, A.S.R.S.: Is NRR Time-Sensitive in Measuring Population Replacement Level, Presented in session 83. Data and methods: A medley of perspectives. In: International Population Conference, Hyderabad, India (2021), IPC 2021. https://ipc2021.popconf.org/abstracts/210503
15. Rao, A.S.R.S.: A partition theorem for a randomly selected large population. Acta Biotheoretica **70**(6), 1–11 (2022). https://doi.org/10.1007/s10441-021-09433-z
16. Rao, A.S.R.S.: A note on geometric structures of stationary populations (unpublished) (2022)
17. Rao, A.S.R.S.: Geometrical Constructions and Analysis in Population Capture Cohorts. Mathematical Population Studies (accepted). Taylor & Francis (2024)
18. Li, N., Tuljapurkar, S.: Population momentum for gradual demographic transitions. Popul. Stud. **53**, 255–262 (1999)
19. Rao, A.S.R.S., Carey, J.R.: Generalization of Carey's equality and a theorem on stationary population. J. Math. Biol. **71**, 583–594 (2015). Springer
20. Carey, J.R., Silverman, S., Rao, A.S.R.S.: The life table population identity: discovery, formulations, proof, extensions, and applications. In: Handbook of Statistics, Vol. 39, pp. 155–186. Elsevier (2018)
21. Rao, A.S.R.S., Carey, J.R.: On the three properties of stationary populations and knotting with non-stationary populations. Bull. Math. Biol. **81**, 4233–4250 (2019). Springer
22. Rao, A.S.R.S., Carey, J.R.: Behavior of stationary population identity in two-dimensions: age and proportion of population truncated in follow-up. In: Handbook of Statistics, Vol. 40, pp. 487–500. Elsevier (2019)
23. Rao, A.S.R.S., Carey, J.R.: Stationary status of discrete and continuous age-structured population models. Math. Biosci. **364**, 109058 (2023). Elsevier
24. Mitra, S.: Influence of instantaneous fertility decline to replacement level on population growth. Demography **13**, 513–519 (1976)
25. Brouard, N.: Mouvements et modeles de population. Institut de formation et de recherches démographiques, Paris, France (1989)
26. Carey, J.R.: Aging in the wild, residual demography and discovery of a stationary population identity. In Sears, R., Lee, R., Burger, O. (eds.) Human Evolutionary Demography. Open Book Publishers, Cambridge, UK (2024)
27. Vaupel, J.W.: Life lived and left: Carey's Equality. Demographic Res. **20**, 7–10 (2009)

28. Age Tables Behind Bars by Ben Polletta: Math Digest of the American Mathematical Society (2014). https://web.archive.org/web/20141029155720/, https://www.ams.org/news/math-in-the-media/mathdigest-index#201410-populations
29. Science News: New theorem determines the age distribution of populations from fruit flies to humans (2014). https://esciencenews.com/articles/2014/10/06/
30. Villavicencio, F., Riffe, T.: Symmetries between life lived and left in finite stationary populations. Demographic Res. **35**, 381–398 (2016)
31. Wrigley-Field, E., Feehan, D.: In a stationary population, the average lifespan of the living is a length-biased life expectancy. Demography **59**(1), 207–220 (2022)
32. Kim, Y.J., Aron, J.L.: On the equality of average age and average expectation of remaining life in a stationary population. SIAM Rev. **31**, 110–113 (1989)
33. Lahiri, S.: Survival probabilities from 5-year cumulative life table survival ratios (Tx + 5/Tx): some innovative methodological investigations. In: Rao, C.R. (ed.) Handbook of Statistics: Integrated Population Biology and Modelling, pp 481–542. Elsevier, Amsterdam (2018)
34. Hanagal, D.D.: Modeling Survival Data Using Frailty Models, xx+314 pp. CRC Press, Boca Raton, FL (2011)
35. Dodd, E., Forster, J.J., Bijak, J., Smith, P.W.F.: Stochastic modelling and projection of mortality improvements using a hybrid parametric/semi-parametric age-period-cohort model. Scand. Actuar. J. **2021**(2), 134–155 (2021)
36. Tuljapurkar, S.: The final inequality: variance in age at death. In: Demography and the Economy, pp. 209–221. University of Chicago Press (2010)
37. Tuljapurkar, S., Edwards, R.: Variance in death and its implications for modeling and forecasting mortality. Demographic Res. **24**, 497–526 (2011)
38. Cox, D.R., Oakes, D.: Analysis of survival data. Monographs on Statistics and Applied Probability. Chapman & Hall, London (1984)
39. Goldstein, J.R.: Life lived equals life left in stationary populations. Demographic Res. **20**(Article 2), 3–6 (2009)
40. Tasnin, M.S., Bode, M., Merkel, K., Clarke, A.R.: A polyphagous, tropical insect herbivore shows strong seasonality in age-structure and longevity independent of temperature and host availability. Sci. Rep. **11**(1), 11410 (2021)
41. Swanson, D.A., Tedrow, L.M.: Two new mathematical equalities in the life table. Canad. Stud. Popul. **49**(2), 67–73 (2021)
42. Carey, J.R., Eriksen, B., Rao, A.S.R.S.: Government as population: Demographic perspectives on the United States legislative, judicial and executive branches, 1789–2020. Model. Anal. Longitudinal Data **50**, 225 (2024)
43. Vaupel, J.W., Villavicencio, F.: Life lived and left: Estimating age-specific survival in stable populations with unknown ages. Demographic Res. **39**, 991–1008 (2018)

Chapter 6
Markov Process Models in Demography

We introduced the concept of a random variable in Chap. 5. Let X be a random variable describing an outcome of an experiment, for example, the outcome of the tossing of a standard coin or a standard dice. The set of all outcomes of a coin tossing experiment is represented by $S = \{Head, Tail\}$, and the dice tossing experiment is represented by $S = \{1, 2, 3, 4, 5, 6\}$. These two are examples of a discrete random variable. Here S is called the sample space. Instead, we could define X as a continuous random variable describing the temperature recorded within an interval of time during a day. The definition of a random variable and its basic properties are explained in Appendix A. Instead of associating a random variable with one experimental outcome, let us measure the outcomes of a random variable repeatedly over time. Let us denote these sequence of random variables by X_n, where n is the discrete time step of an experiment for $n = 0, 1, 2, \ldots$, or let us denote a random variable by $X(t)$ if time is considered as continuous. A *stochastic process* is a collection of X_n or $X(t)$ that evolve over time.

6.1 Markov Chains: Foundations

The Markov process is one kind of stochastic process that we will define soon, and Markov processes arise in many situations in the real world including demography. The values a random variable picks or chooses could be discrete or continuous, a fact we will understand soon in the chapter. The value that a random variable chooses is called the state of the process, and the set of all such states is called the *state space*.

Let $(X_n)_{n \geq 0}$ be a sequence of random variables defined on a state space S. For a moment, let S be discrete. Let X_0 be the initial state of the process. If X_n picks a value i in S at $n = k$ (i.e., $X_n = i$ at $n = k$), then X_{n+1} could stay at i or X_{n+1}

A. S. R. Srinivasa Rao, *Mathematical Demography: Theory and Modeling*,
Mathematical Marvels: Texts and Monographs in the Spirit of CR Rao,
https://doi.org/10.1007/978-981-95-6129-2_6

could move to a different state j in S, that is, $X_{n+1} = j\ (j \neq i)$. A discrete Markov chain is defined as follows:

Definition 6.1 (Discrete Markov Chain) A sequence $(X_n)_{n\geq 0}$ of random variables choosing values in S is called a *discrete Markov chain on* S if for any $n \geq 0$ and any states $i_0, i_1, \ldots, i_n, i, j \in S$, it follows that

$$P\left(X_{n+1} = j / X_0 = i_0, X_1 = i_1, \ldots, X_{n-1} = i_{n-1}, X_n = i\right) = P\left(X_{n+1} = j / X_n = i\right). \tag{6.1}$$

The left side of the equation (6.1)

$$P\left(X_{n+1} = j / X_0 = i_0, X_1 = i_1, \ldots, X_{n-1} = i_{n-1}, X_n = i\right)$$

is a conditional joint probability mass function describing probability of X_{n+1} picking a value (state) j in S given that the joint mass function of random variables $X_0, X_1, \ldots, X_n$ choosing the states $i_0, i_1, \ldots, i_n$, respectively. The right side of equation (6.1)

$$P\left(X_{n+1} = j / X_n = i\right)$$

is an ordinary conditional probability that X_{n+1} is at state j given that X_n is at state i. In the conditional joint function, it took n-time steps to reach the state i from the initial state i_0 and then it took one more time step to reach the state j (from state i). We could interpret that as, on the left side of (6.1) it took $n + 1$ steps to reach the state j from the initial state of i_0 and in the right side of (6.1), it took only one time time step to reach the state j without the requirement of the joint mass function:

$$P\left(X_0 = i_0, X_1 = i_1, \ldots, X_{n-1} = i_{n-1}, X_n = i\right).$$

Equation (6.1) is read as probability that X_{n+1} is at j given that X_0 was at i_0, X_1 was at i_1, X_2 was at i_2, and so on, and X_n was at i is equal to probability that X_{n+1} is at j given that X_n was at i. Since it took one time step to reach j from i, we write

$$P\left(X_{n+1} = j / X_n = i\right) = p_{ij}^{(1)} \text{ or } p_{ij}, \tag{6.2}$$

where p_{ij} is usually read as probability of transition from state i to state j in one step. We use the word *transition* because the X_n was at state i and the random variable moved to the state j at the next step. If $p_{ii} = 1$, we call the state i an *absorbing state*. Similarly, we represent $p_{ij}^{(n)}$ for n-step transitions between any two states in S given by

$$p_{ij}^{(n)} = P\left(X_n = j / X_0 = i\right), \text{ or} \tag{6.3}$$

$$p_{ij}^{(n)} = P\left(X_{n+1} = j / X_1 = i\right), \text{or}$$

Fig. 6.1 The transition probability matrix for the pregnancy cycle Example 6.2

$$P = \begin{array}{c} \\ a \\ b \\ c \end{array}\!\!\begin{array}{c} \begin{array}{ccc} a & b & c \end{array} \\ \begin{pmatrix} p_{aa} & p_{ab} & p_{ac} \\ 0 & p_{bb} & p_{bc} \\ 0 & 0 & p_{cc} \end{pmatrix} \end{array}$$

$$p_{ij}^{(n)} = P\left(X_{n+2} = j/X_2 = i\right), \text{ or,}$$
$$\vdots$$
$$p_{ij}^{(n)} = P\left(X_{n+m} = j/X_m = i\right), \text{ or}$$
$$\vdots \qquad (6.4)$$

Example 6.2 (Pregnancy Cycle) Let X_n be the state of a woman at time n described on a state space $S = \{a, b, c\}$, where a is the nonpregnant state, b is the pregnant state, and c is delivered a baby or became a mother. Let $X_0 = a$. The transition probability $p_{ab}^{(n)}$ represents the probability that a woman who was at state a took n-time steps to reach state b or we could interpret that as the probability that a woman who was not pregnant became pregnant in n-time steps or during n-months if each time step is considered as one month. Similarly, $p_{bc}^{(n)}$ represents the probability that a woman who was pregnant say at $X_m = b$ delivered a baby at $X_{n+m} = c$. These two transition probabilities can be expressed as

$$p_{ab}^{(n)} = P\left(X_n = b/X_0 = a\right) =$$
$$P\left(X_n = b/X_0 = a, X_1 = a, \ldots, X_{n-1} = a\right),$$

and

$$p_{bc}^{(n)} = P\left(X_{n+m} = c/X_m = b\right)$$
$$= P\left(X_{n+m} = c/X_m = b, X_{m+1} = b, \ldots, X_{n+m-1} = b\right).$$

We assume $p_{ba} = p_{ca} = p_{cb} = 0$. The transition probability matrix P for all possible transition between states of S for the pregnancy cycle example is in Fig. 6.1.

Remark 6.3 If there is a positive probability of transition between any two states in S either directly or indirectly through one intermediate state or more than one intermediate states, we call the Markov chain irreducible, and if not, we call the chain reducible. If each row sum of any transition matrix P is equal to one, then we call P a *stochastic matrix*, and if each column sum of P is also equal to one, then we call P a *doubly stochastic*.

Remark 6.4 If a chain leaves a state, say i in S, and there is a positive probability in returning to i only a finite number of times, but eventually the chain leaves i and has a probability of 1 for not to return the state i, we call i a *transient state* . Otherwise, i is called a *recurrent state*.

The $n + m$ step transition probabilities can be written in an elegant form known as the *Chapman–Kolmogorov equations*. These are stated and proved below:

Remark 6.5 (The Chapman–Kolmogorov Equations) Let S be a discrete state space. Then, for a Markov chain

$$p_{ij}^{(n+m)} = \sum_{k \in S} p_{ik}^{(n)} p_{kj}^{(m)}. \tag{6.5}$$

Proof Suppose $i, j, k \in S$. We can write $p_{ij}^{(n+m)}$ as

$$\begin{aligned} p_{ij}^{(n+m)} &= P\left(X_{n+m} = j / X_0 = i\right), \\ &= \sum_{k \in S} P\left(X_{n+m} = j \cap X_n = k / X_0 = i\right), \end{aligned} \tag{6.6}$$

where the right-hand side of (6.6) was obtained through the law of total probability.[1] Now Eq. (6.6) can be written as

$$\begin{aligned} p_{ij}^{(n+m)} &= \sum_{k \in S} P\left(X_{n+m} = j / X_0 = i \cap X_n = k\right) P\left(X_n = k / X_0 = i\right) \\ &= \sum_{k \in S} P\left(X_{n+m} = j / X_n = k\right) P\left(X_n = k / X_0 = i\right) \\ &= \sum_{k \in S} p_{ik}^{(n)} p_{kj}^{(m)}. \end{aligned}$$

□

Remark 6.6 In Example 6.2, it could also happen that $p_{ac}^{(40)} = p_{ab}^{(1)} + p_{bc}^{(39)}$, $p_{ac}^{(55)} = p_{ab}^{(15)} + p_{bc}^{(40)}$, etc.

Definition 6.7 A Markov chain $(X_n)_{n \geq 0}$ is known as *time-homogeneous* if the process is independent of time or independent of time steps, that is,

$$p_{ij} = P\left(X_{n+1} = j / X_n = i\right) = P\left(X_1 = j / X_0 = i\right).$$

[1] If the sample space S is partitioned into n mutually exclusive events $E_1, E_2, \ldots, E_n$ such that $S = E_1 \cup E_2 \cup \ldots \cup E_n$. Let $A = (A \cap E_1) \cup (A \cap E_2) \cup \ldots \cup (A \cap E_n)$. Then, $P(A) = \sum_{i=i}^{n} P(A \cap E_i)$.

Next, we define a continuous-time Markov chain which will help us to build the *birth–death process.*

Definition 6.8 (Continuous-Time Markov Chain) Let $X(t)$ for $t \geq 0$ be a process with a state space S. Then, $X(t)$ for $t \geq 0$ is said to be a *continuous-time Markov chain*, if $A \subset S$, and for

$$s_1 < s_2 < \ldots < s_n < s < t$$

it follows that

$$\begin{aligned} P\left(X(t) \in A/X(s_1) = x_1, X(s_2) = x_2, \ldots, X(s_n) = x_n, X(s) = x\right) \\ = P\left(X(t) \in A/X(s) = x\right), \end{aligned}$$

where $x, x_1, x_2, \ldots, x_n$ are some collection of states in S, i.e.,

$$x, x_1, x_2, \ldots, x_n \in S.$$

6.2 Birth and Death Process

A birth and death model is a classic continuous-time discrete space Markov chain model with state space $S = \{0, 1, 2, \ldots\}$. For any stochastic model, it is helpful to understand the transition probability matrix as we saw above in the pregnancy cycle example. Let i be a state in S, i.e., $i \in S$. Let b_i and d_i be the birth rate and death rate, respectively, at state i for $i = 1, 2, \ldots$. We saw previously the definition for a continuous-time Markov chain.

Let $X(t)$ be a continuous-time Markov chain model describing the births and deaths on S. The following are the three axioms describing the process:

Axiom I. The probability of having one birth at state i during time interval of length Δt, say during $[t, t + \Delta t]$, is represented by $b_i \Delta t + o(\Delta t)$.[2] That is, if the population size is i at time t, then the population size increases to $i + 1$ with one birth during $[t, t + \Delta t]$. We express this by

$$p_{i,i+1}(\Delta t) = P\{X(t + \Delta t) = i + 1/X(t) = i\} = b_i \Delta t + o(\Delta t). \tag{6.7}$$

From (6.7), we can express the probability of having no birth during $[t, t + \Delta t]$ as $1 - b_i \Delta t + o(\Delta t)$.

[2] Here $o(\Delta t)$ means a small change in time.

Axiom II. The probability of one death at state i during $[t, t + \Delta t]$ is represented by $d_i \Delta t + o(\Delta t)$. That is, if the population size is i at time t, then the population size decreases to $i - 1$ with one death during $[t, t + \Delta t]$. We express this by

$$p_{i,i-1}(\Delta t) = P\{X(t + \Delta t) = i - 1/X(t) = i\} = d_i \Delta t + o(\Delta t). \tag{6.8}$$

If the population size is 0, then there is no chance of death, i.e. $d_0 = 0$.

Axiom III. The probability of having no birth or no death during $[t, t + \Delta t]$ is represented by $1 - (b_i + d_i)\Delta t + o(\Delta t)$. We express this by

$$p_{i,i}(\Delta t) = P\{X(t + \Delta t) = i/X(t) = i\} = 1 - (b_i + d_i)\Delta t + o(\Delta t). \tag{6.9}$$

For example, if the population size is 1 at time t, then at time $t + \Delta t$, the population size could increase to 2 (1 birth during $[t, t + \Delta t]$) with probability $b_1 \Delta t + o(\Delta t)$, or the population size could decrease to 0 (1 death during $[t, t + \Delta t]$) with probability $d_1 \Delta t + o(\Delta t)$ or the population size could remain the same during $[t, t + \Delta t]$ without a birth or a death with probability $1 - (b_1 + d_1)\Delta t + o(\Delta t)$.

For example, if the population size is 15 at time t, then at time $t + \Delta t$, the population size could increase to 16 (1 birth during $[t, t + \Delta t]$) with probability $b_{15}\Delta t + o(\Delta t)$, or the population size could decrease to 14 (1 death during $[t, t + \Delta t]$) with probability $d_{15}\Delta t + o(\Delta t)$ or the population size could remain at 15 during $[t, t + \Delta t]$ without a birth or a death with probability $1 - (b_{15} + d_{15})\Delta t + o(\Delta t)$.

We can similarly write other possibilities if the population size is N.

The birth and death process also assumes that the probability of more than one event during $[t, t + \Delta t]$ is zero. That the probability of occurring an event of more than one birth, an event of more than one death, or events of one birth and one death during $[t, t + \Delta t]$ is assumed to be zero. Next, our aim is to obtain an expression for the probability function $P_i(t + \Delta t)$ describing there are i individuals at time $t + \Delta t$ (population size at $t + \Delta t$ is i), and then we obtain an expression for the differential equation $P_i'(t)$.

Let $P_i(t + \Delta t)$ be given for $i > 0$. There are three possible following mutually exclusive events[3] under this situation:

E_1 : the population size is i at t and there are no births and no deaths during $[t, t + \Delta t]$.
E_2 : the population size is $i - 1$ at t and there is one birth during $[t, t + \Delta t]$.
E_3 : the population size is $i + 1$ at t and there is one death during $[t, t + \Delta t]$.

[3] Two events E_1 and E_2 are mutually exclusive or disjoint if the probability of occurrence of E_1 and E_2 together is zero, i.e., $P(E_1 \cap E_2) = 0. P(E_1 \cup E_2) = P(E_1) + P(E_2) - P(E_1 \cap E_2) = P(E_1) + P(E_2)$.

Only one out of E_1, E_2, or E_3 can occur during $[t, t + \Delta t]$. The corresponding probabilities of occurrence of these three events can be expressed as

$$\begin{aligned} P(E_1) &= P_i(t)\,[1 - (b_i + d_i)\Delta t + o(\Delta t)]\,, \\ P(E_2) &= P_{i-1}(t)\,[b_{i-1}\Delta t + o(\Delta t)]\,, \\ P(E_3) &= P_{i+1}(t)\,[d_{i+1}\Delta t + o(\Delta t)]\,. \end{aligned} \tag{6.10}$$

Given the possibility of occurrence of one of the events E_1, E_2, or E_3, we write $P_i(t + \Delta t)$ using (6.10) as[4]

$$\begin{aligned} P_i(t + \Delta t) &= P(E_1) + P(E_2) + P(E_3) \\ &= P_i(t)\,[1 - (b_i + d_i)\Delta t + o(\Delta t)] + P_{i-1}(t)\,[b_{i-1}\Delta t + o(\Delta t)] + \\ &\quad P_{i+1}(t)\,[d_{i+1}\Delta t + o(\Delta t)] \\ &= P_i(t)\,[1 - (b_i + d_i)\Delta t] + P_{i-1}(t)b_{i-1}\Delta t + \\ &\quad P_{i+1}(t)d_{i+1}\Delta t + o(\Delta t). \end{aligned} \tag{6.11}$$

We rearrange the terms in (6.11) further and ignore $o(\Delta t)$ to obtain

$$\begin{aligned} P_i(t + \Delta t) - P_i(t) = -P_i(t)(b_i + d_i)\Delta t + P_{i-1}(t)b_{i-1}\Delta t + \\ P_{i+1}(t)d_{i+1}\Delta t. \end{aligned} \tag{6.12}$$

Dividing both sides of (6.12) by Δt and taking $\Delta t \to 0$, we get

$$\begin{aligned} \lim_{\Delta t \to 0} \frac{P_i(t + \Delta t) - P_i(t)}{\Delta t} = -P_i(t)(b_i + d_i) + P_{i-1}(t)b_{i-1} + \\ P_{i+1}(t)d_{i+1}. \end{aligned} \tag{6.13}$$

From the definition of derivative of a function, the left side of (6.13) is written as $P_i'(t)$; thus,

$$P_i'(t) = -P_i(t)(b_i + d_i) + b_{i-1}P_{i-1}(t) + d_{i+1}P_{i+1}(t) \text{ for } i > 0. \tag{6.14}$$

Let $P_0(t + \Delta t)$ be given. That is, the population size at $t + \Delta t$ is 0. For this situation, there is a possibility of occurring of one of the two following mutually exclusive events:

E_4 : the population size is 0 at t and there are no births during $[t, t + \Delta t]$ (as noted earlier, when the population size is 0, then $d_0 = 0$).
E_5 : the population size is 1 at t and there is one death during $[t, t + \Delta t]$.

[4] $P_i(t + \Delta t) = P(E_1 \text{ or } E_2 \text{ or } E_3) = P(E_1) \cup P(E_2) \cup P(E_3)$.

The corresponding probabilities for E_4 and E_5 are expressed as

$$P(E_4) = P_0(t)\,[1 - b_0\Delta t + o(\Delta t)]\,,$$
$$P(E_5) = P_1(t)\,[d_1\Delta t + o(\Delta t)]\,. \tag{6.15}$$

We write $P_i(t + \Delta t)$ using (6.15) as

$$P_0(t + \Delta t) = P_0(t)\,[1 - b_0\Delta t] + P_1(t)\,[d_1\Delta t] + o(\Delta t). \tag{6.16}$$

Following the steps that led to (6.14), we obtain $P_0'(t)$ as

$$P_0'(t) = -b_0 P_0(t) + d_1 P_1(t). \tag{6.17}$$

The two equations (6.14) and (6.17) together are known as the birth and death differential equation model for the Markov chain $X(t)$ described above and are also known as *forward Kolmogorov differential equations*. In this model, $P_i(t)$ are variables, and b_i and d_i for $i = 0, 1, 2, ..$ are parameters. For such differential equations, we obtain a steady-state solution or local stability solutions for $P_i(t)$ for $i \geq 0$ by equating $P_i'(t)$ in (6.14) to *zero* and $P_0'(t)$ in (6.17) to zero.[5] In the steady state, the system becomes independent of time, so we do not have to mention t in $P_i(t)$. When $P_0'(t) = 0$ in (6.17), we get

$$P_1 = \frac{b_0}{d_1} P_0. \tag{6.18}$$

When $P_i'(t) = 0$ in (6.14), we get

$$-P_i(t)(b_i + d_i) + b_{i-1}P_{i-1}(t) + d_{i+1}P_{i+1}(t) = 0 \text{ for } i > 0. \tag{6.19}$$

By substituting $i = 1$ in (6.19) and simplifying further, we get

$$P_2 = \frac{(b_1 + d_1)P_1 - b_0 P_0}{d_2} = \frac{b_0 b_1}{d_1 d_2} P_0 \text{ (we substituted } P_1 \text{ from (6.18))}\,.$$

By substituting $i = 2$ in (6.19), and simplifying further, we get

$$P_2 = \frac{b_0 b_1 b_2}{d_1 d_2 d_3} P_0,$$

[5] Suppose $\frac{dX}{dt} = f_1(X, Y)$ and $\frac{dY}{dt} = f_2(X, Y)$ are two differential equations with two variables X and Y. Under the steady state, we solve for X and Y by making $\frac{dX}{dt} = 0$ and $\frac{dY}{dt} = 0$. That is, we solve X and Y by making $f_1(X, Y) = 0$ and $f_2(X, Y) = 0$. Steady-state solutions of population dynamics or ecology models were described in detail in Chap. 4.

and continuing further in a similar way for $i = 1, 2, 3, \ldots$ and simplifying further, we arrive at

$$P_i = \prod_{j=0}^{i-1} \frac{b_j}{d_{j+1}} P_0 \text{ for } i = 1, 2, 3, \ldots \quad (6.20)$$

Similar to the two differential equations (6.14) and (6.17), we could also derive the two following differential equations:

$$P_0'(t) = -\, b_0 P_0(t) + b_0 P_1(t), \quad (6.21)$$

$$P_i'(t) = -\, P_i(t)(b_i + d_i) + b_i P_{i+1}(t) + d_i P_{i-1}(t) \text{ for } i > 0. \quad (6.22)$$

Equations (6.21) and (6.22) are known as *backward Kolmogorov differential equations* (Fig. 6.2).

The birth and death process provides a useful intuition on how continuous-time stochastic models work, especially under the Markov property framework (Fig. 6.3). Due to the assumption of not more than one event within the interval $[t, t + \Delta t]$, the random variable jumps to either one state higher or one state lower (except for the state 0) or remains in the same state if there are no birth and deaths at a given state. For further discussions and other restrictions on such processes, refer to [1–4]. Not only in demography, the birth and death process framework plays interesting roles in various other fields, and these are easy to model other kinds of population dynamics

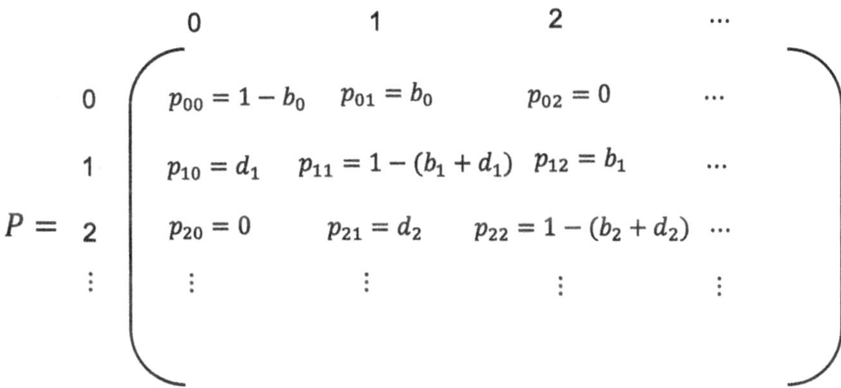

Fig. 6.2 The transition probability matrix for the birth and death process. The state space S of a standard birth and death process is given by $S = \{0, 1, 2, \ldots\}$. The quantities b_i for $i = 0, 1, 2, \ldots$, and d_i for $i = 1, 2, \ldots$ are the probability of having a birth at state i and the probability of a death at state i, respectively. When $b_i = b$ and $d_i = 0$, it becomes a pure birth process or also referred as a Poisson process

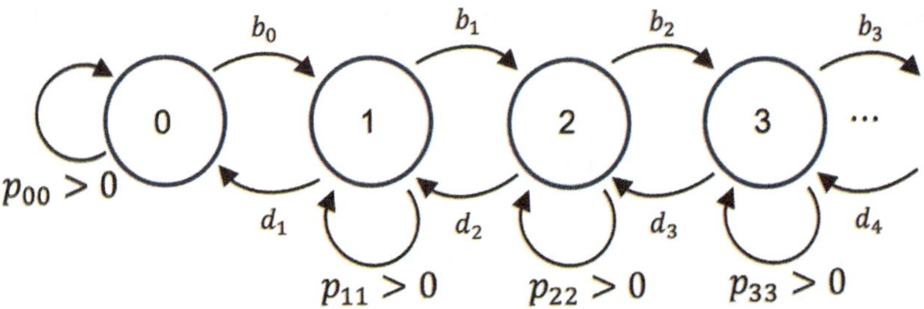

Fig. 6.3 Transition graph of birth and death process. The numbers 0, 1, 2, ... indicate the states in the state space S, b_i is chance of birth at state i for $i = 0, 1, 2, \ldots$, d_i is chance of death at state i for $i = 1, 2, 3, \ldots$. The quantity $p_{ii} > 0$ indicates that there is a positive probability for $X(t)$ to remain at i for $i = 0, 1, 2, \ldots$

[5–14]. When $b_i = b$ (i.e., constant birth rate across all the states in S) and $d_i = 0$, then the birth and death process becomes a Poisson process and is also sometimes called a *pure birth process*.[6] For further references on Markov chain applications in demography, refer to [15–17]. Next, we will demonstrate other demography-related applications of Markov chains, namely, infertility studies.

[6] A pure birth process can be derived from the differential equations $P_i'(t)$ for $i > 1$ and $P_0'(t)$ seen in the birth death process. When $b_i = b$ and $d_i = 0$ in (6.14) and (6.17), we arrive at

$$P_i'(t) = -bP_i(t) + bP_{i-1}(t) \text{ for } i > 0, \tag{6.23}$$

and

$$P_0'(t) = -bP_0(t). \tag{6.24}$$

Equation (6.24) is a Malthus-type ordinary differential equation and the solution of such equation we saw in (2.22) and (2.23). The solution of (6.24) is

$$P_0(t) = e^{-bt}. \tag{6.25}$$

Next, we use probability generating function (p.g.f.) to solve (6.23). Such functions are also useful in the analysis of branching process described in Chap. 7. See also footnote 2. Let $G_s(t)$ be the p.g.f. of $P_i(t)$ in (6.23), where

$$G_s(t) = \sum_{i=0}^{\infty} s^i P_i(t). \tag{6.26}$$

Multiplying by s^i to both sides of (6.23) and summing over i (from 1 to ∞), we get

$$\sum_{i=1}^{\infty} s^i P_i'(t) = -b \sum_{i=1}^{\infty} s^i P_i(t) + b \sum_{i=1}^{\infty} s^i P_{i-1}(t). \tag{6.27}$$

6.3 Infertility Models

Let us build a discrete-time discrete-space Markov chain model to understand the impacts of treatment options for women who are undergoing treatments like gonadotropin and aromatase inhibitors for infertility. In recent years, Markov models have been built for understanding infertility treatment data [18, 19]. The idea was to compute probabilities of transitions between states (health status) of women undergoing the two kinds of treatments mentioned above.

Let X_n be a random variable describing the infertility treatment, normal pregnancy, ectopic pregnancy,[7] becoming a mother, etc. All the women considered in the study have come to an infertility clinic or an IVF clinic for one or other infertility-related problems that they are suffering, and they are provided a treatment. Let S be the state space describing the state of $(X_n)_{n \geq 0}$. All women were monitored at discrete time points until a baby was delivered (if conceived), and all the women were reinitiated with one of the therapies if not conceived after a certain number of cycles, or a woman could have withdrawn from a study failure to respond to treatment after a prescribed number of treatment cycles. Let $S = \{0, 1, 2, 3, 4\}$ be the state space describing the value of X_n for $n = 0, 1, 2, \ldots, k$ for some k number of finite reproductive pregnancy treatment cycles. The states of S are described as follows:

0 : a woman is infertile and has started treatment.
1 : a woman who has been on treatment has conceived.
2 : a woman who has been on treatment is dropped from the study due to not being successful in conceiving after prescribed treatment cycles.
3 : a conceived woman ends pregnancy due to a miscarriage or stillbirth.
4 : a conceived woman delivers a live baby.

Combining (6.27) and (6.24), we get

$$\sum_{i=0}^{\infty} s^i P_i'(t) = -b \sum_{i=0}^{\infty} s^i P_i(t) + b \sum_{i=1}^{\infty} s^i P_{i-1}(t). \tag{6.28}$$

Applying (6.26) in (6.28), we get

$$\frac{d}{dt} G_s(t) = -bG_s(t) + bsG_s(t) = b(s-1)G_s(t),$$

and the solution of the above differential equation (similar ones we saw earlier in the chapter) is

$$G_s(t) = \exp\{bt(s-1)\}. \tag{6.29}$$

We see that $G_s(t)$ in (6.29) is the probability generating function of the Poisson distribution, confirming that the Poisson process is a pure birth process.

[7] Ectopic pregnancy is referred to here when the fertilized egg is situated outside the uterus and attached to it.

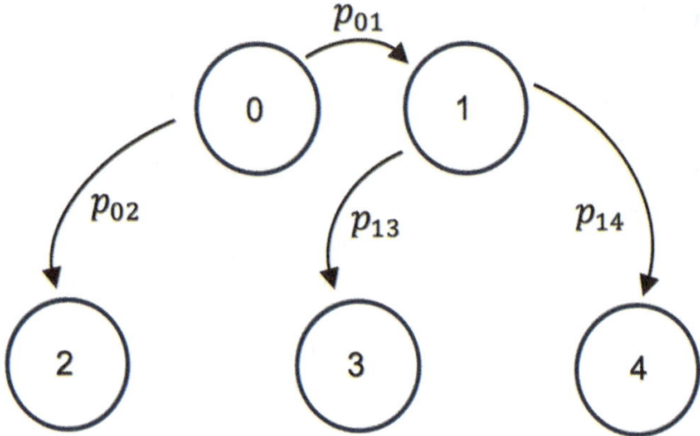

Fig. 6.4 All possible transitions between states of S of infertility treatment. The numbers 0, 1, 2, 3, and 4 are states of the Markov chain, and p_{ij} is the transition probability from state i to state j for $i = 0, 1$, and $j = 1, 2, 3, 4$

Table 6.1 Hypothetical data on two infertility treatments and status of women after completing all treatment cycles

	The number of women in states of S				
Treatment name	0	1	2	3	4
Gonadotropin	512	309	203	31	278
Aromatase inhibitors	496	270	226	40	230

All possible transitions between states are explained in Fig. 6.4. The probability p_{ij} of a transition from state i in S to state j in S is defined as

$$p_{ij} = \frac{\text{the number of women who have moved eventually from } i \text{ to } j}{\text{the number of women who were at state } i}. \tag{6.30}$$

The transition probability matrix P for the process $X(t)$ of infertility treatment can be constructed using the data presented in Table 6.1. Let $p_{ij}^{(G)}$ and $p_{ij}^{(AI)}$ are the transition probabilities between states i and j for gonadotropin (G) and aromatase inhibitors (AI), and $P^{(G)}$ and $P^{(AI)}$ are corresponding transition probability matrices, respectively. We can compare the difference between the impacts of these two treatments using z-scores and standard errors (SEs) using the formulae provided below:

$$z_{ij} = \frac{p_{ij}^{(G)} - p_{ij}^{(AI)}}{SE_{ij}}, \tag{6.31}$$

and the standard errors can be computed using

$$SE_{ij} = \sqrt{\left\{ p_{ij}(1 - p_{ij}) \left(\frac{1}{n_{ij}(G)} + \frac{1}{n_{ij}(AI)} \right) \right\}}, \tag{6.32}$$

where

$$p_{ij} = \frac{p_{ij}^{(G)} n_{ij}(G) + p_{ij}^{(AI)} n_{ij}(AI)}{n_{ij}(G) + n_{ij}(AI)}, \tag{6.33}$$

and $n_{ij}^{(G)}$ is the number of transitions from i to j who are under the treatment gonadotropin, and $n_{ij}^{(AI)}$ is the number of transitions from i to j who are under the treatment aromatase inhibitors. The transition matrices for two treatments based on Table 6.1 are provided in Fig. 6.5. As an example, we have tested the hypothesis

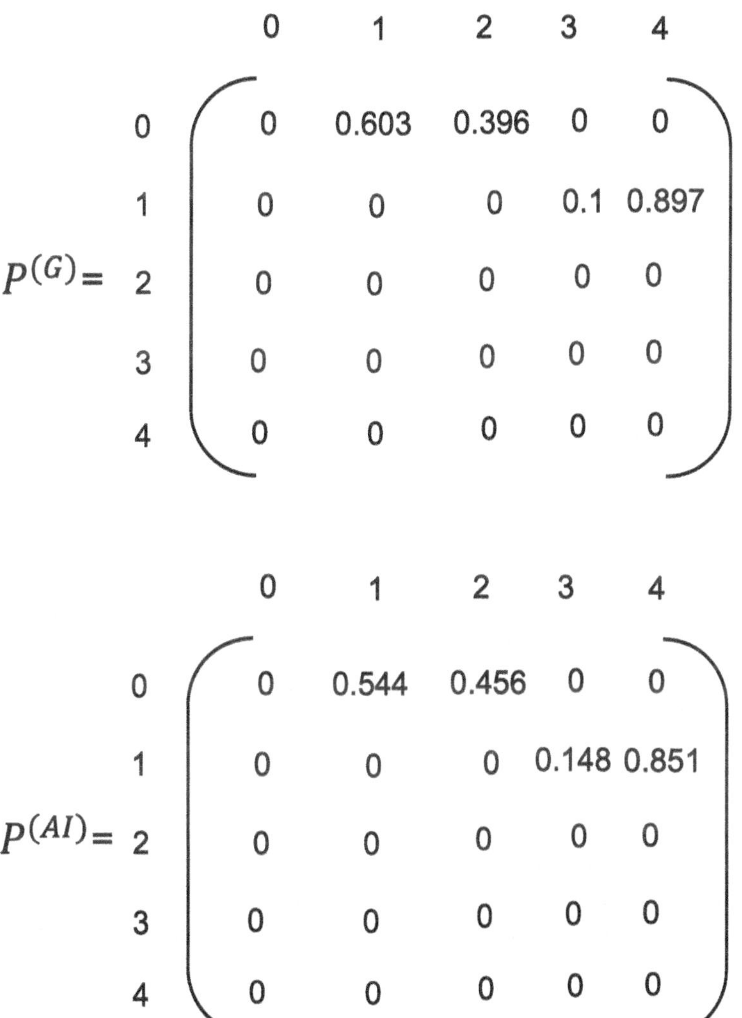

Fig. 6.5 Transition probability matrices for two treatments: gonadotropin and aromatase inhibitors, are computed from the data in Table 6.1

that the treatment impacts in delivering a baby would be equal after conceiving and computed the corresponding z-scores, namely, z_{14} using the formulae (6.31)–(6.33). The computations of p_{14}, SE_{14}, and z_{14} are

$$p_{14} = \frac{(0.897 \times 278) + (0.851 \times 230)}{278 + 230} = 0.876,$$

$$SE_{14} = \sqrt{\left\{0.876(1 - 0.876)\left(\frac{1}{278} + \frac{1}{230}\right)\right\}} = 0.029,$$

and

$$(\text{z-score})\ z_{14} = \frac{0.897 - 0.851}{0.029} = 1.586.$$

Since the value of $z_{14} = 1.586$ is less than 1.96, the critical value obtained from z tables at $z_{1-0.05/2}$, we conclude that there is no significant difference between the two types of treatments. If a computed z_{14} for a different large dataset exceeds 1.96, we conclude that gonadotropin treatment is significantly better than aromatase inhibitor treatment. For a discussion of clinical aspects and Markov models to compare in fertility treatments, refer to Rao and Diamond (2017) [19].

6.4 Deep Learning Models in Infertility

Deep learning models and artificial intelligence models rely on mapping and building mathematical structures, mathematical reasoning on the data, analysis using the data, predictions, and bringing useful conclusions [20, 21]. As the data becomes very large and the dimension is huge, especially in demographic studies, machine learning and deep learning algorithms become inevitable for reducing the time involved in decision-making and timely interventions in the populations. One such example was demonstrated in analyzing the global infertility data and bringing time-sensitive conclusions on the type of treatments that an infertile couple could opt for to increase their chances of having a baby [18]. In this chapter, we describe an outline of a deep learning algorithm developed in [18].

Let $c_1, c_2, \ldots, c_k$ be k- possible clinical classifications; $d_1, d_2, \ldots, d_l$ be l-possible demographic classifications; $g_1, g_2, \ldots, g_m$ be m- possible genetic classifications; and $o_1, o_2, \ldots, o_n$ be n- possible other classifications that infertile women posses that require a treatment. Let Ω be the set of all permutations of these classifications, where

$$\Omega = \{1, 2, \ldots, klmn\}. \tag{6.34}$$

See Appendix D on permutations. Let O be the set of treatment options available for a woman in the population, where $O = \{r_1, r_2, \ldots, r_n\}$. Every infertile woman who received a treatment will have one of the characteristics of the set Ω. We compute the transition probability p_{14} for each woman in the population who underwent a treatment in O and reached the state 1 described in the state space S in Sect. 6.3. Let $p_{14}(\delta, o)$ be the transition probability for a woman with a characteristic δ and underwent a treatment o, where $\delta \in \Omega$ and $o \in O$. We compute the following two quantities:

$$\max_{\delta}\left\{\max_{o}\left(p_{14}(\delta, o)\right)\right\}, \tag{6.35}$$

and

$$\max_{o}\left\{\max_{\delta}\left(p_{14}(\delta, o)\right)\right\}. \tag{6.36}$$

The goal of the deep learning algorithm in this context is to assist women experiencing infertility who visit an IVF clinic in finding the most effective treatment options to achieve pregnancy. The two values, (6.35) and (6.36), derived from extensive global treatment data, will help identify the right combinations of treatments for each individual case. For example, if a woman with a characteristic, say x visits an IVF clinic, where $x \in \Omega$, then, from previously computed global data on IVF treatments and their pregnancy outcomes, the algorithm will compute

$$\max\left\{p_{14}\left(x : r_1, r_2, \ldots, r_n\right)\right\} = r_\theta \text{ (say)}, \tag{6.37}$$

where $p_{14}\left(x : r_1, r_2, \ldots, r_n\right)$ is the transition probabilities for all the women with characteristic x who underwent n-treatment options

$$r_1, r_2, \ldots, r_n.$$

For whichever treatment option in O, we obtained the maximum probability of delivering a baby in (6.37); in this case r_θ, the woman with characteristics x will be recommended with the treatment r_θ.

This kind of deep learning model will help increase the chance of delivering a baby for couples and assist in clinical decision-making. The availability of global data on infertility or any practical situation in which deep learning models are developed is essential before adapting to such algorithms. The advantages and limitations of deep learning and artificial intelligence models can be seen elsewhere [20–22]. The central advantage of deep learning models in such infertility studies is that the treatment success patterns available globally with diversified clinical-genetic and demographic information can be utilized to treat a couple or a woman at an IVF clinic that is serving the population that has come to the clinic from anywhere in the world.

6.5 Exercises

Exercise 6.9 Which of the states in Example 6.2 of the pregnancy cycle are transient? Which of the states are recurrent? Justify.

Exercise 6.10 Let $N(t)$and $N(t + \Delta t)$ be the population sizes at times t and $t + \Delta t$, respectively, and let these two quantities be related through the quantity θ, the growth rate of the population as follows: $N(t + \Delta t) - N(t) = \theta \Delta t N(t)$. Build a stochastic process model on the discrete state space $\{0, 1, 2, \ldots\}$ of population sizes and on continuous time.

Exercise 6.11 For the model in Exercise 6.10, let $X_t = \{0, 1, 2, \ldots\}$ and $P_N(t) = \text{Prob}[X_t = N]$. Let P_0 be the initial population. Assume that for a birth rate b,

$$\text{Prob[one birth during } (t, t + \Delta t)] = b\Delta t$$

and

$$\text{Prob[more than one birth during } (t, t + \Delta t)] \approx 0.$$

Finally, assume that $\text{Prob}[X_0 = P_0] = 1$. Deduce the following differential equation:

$$\frac{dP_N(t)}{dt} = P_{N-1}(t)b(N-1)\Delta t + P_N(t)\left(1 - bN\Delta t\right).$$

The equation mentioned above is commonly known as the *Kolmogorov forward differential equation*.

Exercise 6.12 Let $b_i = b^2 i + M$ (for $M > 0$) and $d_i = d^2 i$ for $i = 0, 1, 2, \ldots$ in a birth and death process $X(t)$ described by the two following differential equations:

$$\frac{P_i(t)}{dt} = -P_i(t)(b_i + d_i) + b_{i-1}P_{i-1}(t) + d_{i+1}P_{i+1}(t) \text{ for } i > 0,$$

$$\frac{P_0(t)}{dt} = -b_0 P_0(t) + d_1 P_1(t).$$

Show that $E(X(t))$ satisfies the initial value problem $\frac{E(X(t)}{dt} = M + (b^2 - d^2)\sum_{i=1}^{\infty} iP_i$ with $E(X(0)) = i$ and $X(0) = i$. Show that the solution of this initial value problem is

$$E\left((t)\right) = \begin{cases} Mt + i \text{ if } b^2 = d^2 \\ \frac{M}{b^2-d^2}\left(e^{(b^2-d^2)t} - 1\right) + ie^{(b^2-d^2)t} \text{ if } b^2 \neq d^2 \end{cases}.$$

Note that i in the expression of $E\left((t)\right)$ above represents the state i.

References

1. Karlin, S.: A First Course in Stochastic Processes. Academic Press, New York, London (1996)
2. Medhi, J.: Stochastic Processes, 2nd edn. Wiley, New York (1994)
3. Goswami, A., Rao, B.V.: A Course in Applied Stochastic Processes. Texts and Readings in Mathematics, vol. 40. Hindustan Book Agency, New Delhi (2006)
4. Lawler, G.F.:. Introduction to Stochastic Processes, 2nd edn. Chapman & Hall/CRC, Boca Raton (2006)
5. Parthasarathy, P.R., Lenin, R.B., Schoutens, W., Van Assche, W.: A birth and death process related to the Rogers–Ramanujan continued fraction. J. Math. Anal. Appl. **224**(2), 297–315 (1998)
6. Sharma, O.P., Nair, N.S.K.: Transient analysis of finite state birth and death process with absorbing boundary states. Stochast. Anal. Appl. **14**(5), 565–589 (1996)
7. Yadava, R.C., Rai, P.K.: Analyzing variety of birth intervals: a stochastic approach. Integrated Population Biology and Modeling. Part B, pp. 195–283. Handbook of Statist., vol. 40. Elsevier/North-Holland, Amsterdam (2019)
8. Singh, K.K., Singh, B.P., Singh, U., Singh, K.: Trends in fecundability and sterility over time. Progr. Math. (Varanasi) **36**(1–2), 111–129 (2002)
9. Rao, A.S.R.S.: Multilevel contours on bundles of complex planes☆. In: Handbook of Statistics, Vol. 46, pp. 401–464. Elsevier (2022)
10. Tuljapurkar, S., Steinsaltz, D.: Stochastic models for structured populations. In: Handbook of Statistics, Vol. 40, pp. 133–155. Elsevier (2019)
11. Sukhija, S., Kumar, R., Rao, A.S.R.S.: Continuous Time Markov Modeling and Matrix Analysis for Infectious Disease Spread. Sankhya B. Springer (2025)
12. Sukhija, S., Kumar, R., Rao, A.S.R.S.: Multi-State Semi-Markov Model for Chikungunya Infection Progression with Sensitivity and Elasticity Measures (unpublished)
13. Singh, B.P., Singh, S., Das, U.D.: Development of Stochastic Models for Waiting time to First Conception through the Nature of Hazard. Sankhya B. Springer (2025)
14. Rao, C.R., Rao, A.S.R.S. (eds.): Markov Chains: Theory and Applications, Handbook of Statistics, vol. 52. Elsevier (2025)
15. Pathak, K.B., Pandey, A.: Stochastic Models of Human Reproduction, by K.B. Pathak, A. Pandey. Himalaya Publishing House (1993)
16. Pasupuleti, S.S.R., Chattopadhyay, A.K.: Probability distributions of number of children and maternal age at various order births using age-specific fertility rates by birth order. Sankhya B **75**(2), 374–408 (2013)
17. Caswell, H.: Applications of Markov chains in demography. In: Langville, A.N., Stewart, W.J. (eds.). MAM2006: Markov Anniversary Meeting. Raleigh, pp. 319–334. Boson Books, North Carolina (2006)
18. Rao, A.S.R.S., Diamond, M.P.: Deep learning of Markov model-based machines for determination of better treatment option decisions for infertile women. Reproduct. Sci. **27**, 763–770 (2020)
19. Rao, A.S.R.S., Diamond, M.P.: Role of Markov modeling approaches to understand the impact of infertility treatments. Reproduct. Sci. **24**(11), 1538–1543 (2017)
20. Govindaraju, V., Rao, A.S.R.S., Rao, C.R. (Eds.): Deep Learning. Elsevier (2023)
21. Krantz, S.G., Rao, A.S.R.S., Rao, C.R. (Eds.): Artificial Intelligence. Elsevier (2023)
22. Rao, A.S.R.S.: Exact deep learning machines. In: Govindaraju, V., Rao, A.S.R.S., Rao, C.R. (eds.) Handbook of Statistics, vol. 48, pp. 1–8. Elsevier (2023)

Chapter 7
Branching Process

The theory of the *branching process* (also a part of the stochastic process) was developed with the aim of understanding the extinction probabilities of family surnames that transfer between generations through male offspring. Males carry their family's surnames in England and Francis Galton wanted to study the probabilities of extinction of family surnames in medieval England based on the number of male offspring that a couple has in each generation. The field of branching process could have originated from studies on son preference in demography, where financial assets and surnames were passed down through sons in medieval England. In this chapter, we first start with the formulation of the branching process framework. These basics emerge from the convergence principles of mathematical analysis and probability functions. Later in the chapter, we state and prove a classical probability extinction theorem. The branching process model has been fascinating since the development in the late nineteenth century.

In continuation of the Markov modeling techniques introduced in the previous chapter, we now explore additional stochastic modeling and analysis techniques, specifically, branching processes. Together, these two chapters provide a solid foundation for students to understand and apply stochastic process models in demography. The integration of Markov chains and branching processes equips learners with essential tools to analyze population dynamics, generational growth, and probabilistic transitions in demographic systems.

7.1 Model Formulation

Let $(X_n)_{n\geq 0}$ be a stochastic process describing the size of the male population in the nth generation. Later we will see why $(X_n)_{n\geq 0}$ can be called a branching process. Although we will study extinction probabilities here, the concept of the branching process is related to the uniform convergence of a sequence of functions

A. S. R. Srinivasa Rao, *Mathematical Demography: Theory and Modeling*,
Mathematical Marvels: Texts and Monographs in the Spirit of CR Rao,
https://doi.org/10.1007/978-981-95-6129-2_7

in *mathematical analysis*. Further details on uniform convergence of functions can be found in Appendix E.

A series of the form $\sum_{n=0}^{\infty} a_n x^n$ is called a power series in x. This series is said to converge at x if the limit of

$$\sum_{n=0}^{N} a_n x^n$$

as $N \to \infty$ exists. We came across this type of series in Chap. 6 while trying to understand the pure birth process (also known as Poisson process). See Footnote 6. The power series $G_s(t)$ in Chap. 6 and in Footnote 6 was a probability generating function. We utilize such series in understanding branching process extinction probabilities. If a power series $G(x) = \sum_{n=0}^{\infty} a_n x^n$ converges at a point $x = M > 0$, then the series uniformly converges on the interval $[0, M]$. The function $G(x)$ converges at $x = 0$ and $G(0) = a_0$. The derivative of this series is expressed by

$$G'(x) = a_1 + 2a_2 x + 3a_3 x^2 + \ldots, \tag{7.1}$$

and $G(x)$ is infinitely differentiable on $(-M, M)$. That means $G'(x), G''(x), G'''(x), \ldots$ exists and these derivatives are continuous on $[-M, M]$, where

$$G''(x) = 2a_2 + 3.2a_3 x + 4.3.a_4 x^2 + \ldots,$$

$$G'''(x) = 6a_3 + 24a_4 x + 60a_5 x^2 + \ldots,$$

and so on. Evaluating the functions $G'(x), G''(x), G'''(x), \ldots$ at $x = 0$, we have

$$a_1 = G'(0),\ a_2 = \frac{G''(0)}{2!},\ a_3 = \frac{G'''(0)}{3!}, \ldots.$$

The value of the limit of a power series is called the sum of the power series.

Each male offspring in a generation forms a branch and the branches continue to generate more branches through further male offspring. We will soon see that the extinction probabilities are expressed as terms of a power series. In a generation, let p_0 be the proportion of couples with no male offspring, p_1 be the proportion of couples with one male offspring, and, continuing, let p_k be the proportion of couples with k number of male offspring, and so on. If there are no male offspring after k generations, then the surname associated with the couple will extinct. However, if $p_0 = 0$, that is, in a generation, every couple has a male offspring, then the surnames will never extinct. Let θ_k be the probability of extinction of a given surname after k generations. When $p_0 = 0$, obviously $\theta_k = 0$ for $k = 0, 1, 2, \ldots$ because every family or a couple has at least one male in every generation. To make the question interesting, we assume $p_0 > 0$. This implies $\theta_1 = p_0$.

Assume p_k for $k = 0, 1, 2, \ldots$ is constant over all the generations. Then

$$\theta_1 \leq \theta_2 \leq \theta_3 \leq \ldots . \tag{7.2}$$

Since $(\theta_k)_{k\geq 1}$ is bounded because $0 \leq \theta_k \leq 1$, the sequence θ_k in (7.2) is monotone. By monotone convergence theorem,[1] the sequence θ_k converges. Let (θ_k) converge to θ (say). That is,

$$\lim_{k\to\infty} \theta_k = \theta. \tag{7.3}$$

Let us express the power series function $G(x)$ at the point $x = \theta_k$ with the coefficients $p_0, p_1, p_2, \ldots$ by replacing the existing coefficients $a_0, a_1, a_2, \ldots$

$$G(\theta_k) = p_0 + p_1\theta_k + p_2\theta_k^2 + \ldots . \tag{7.4}$$

In (7.4), the term p_0 is the proportion of couples in a generation with no male, the term $p_1\theta_k$ describes the proportion of couples with one male offspring and they will extinct after k generations, the term $p_2\theta_k^2$ describes the proportion couples with two male offspring and the surname tree of each male offspring will extinct after k generations. Similarly, the term $p_j\theta_k^j$ describes the proportion couples with j male offspring and surname tree of each male offspring will extinct after k generations, and so on for other higher-order terms in the power series. This gives us

$$G(\theta_k) = \theta_{k+1}. \tag{7.5}$$

Three observations can be made using (7.4).

Observation I: In (7.4), the terms p_n for $n = 0, 1, 2, \ldots$ can be treated as the sequence $(p_n)_{n\geq 0}$ or the probability mass function of a random variable, say X, because $\Sigma_{n=0}^{\infty} p_n = 1$. That means $G(\theta_k)$ is the probability generating function[2] of X. See Appendix B for a definition of probability mass function (p.m.f.).

Observation II: When $\theta_k = 0$, we have $G(0) = p_0$, and when $\theta_k = 1$, we have

$$G(1) = p_0 + p_1 + p_2 \ldots = \Sigma_{n=0}^{\infty} p_n = 1 \text{ (since the sequence } p_n \text{ forms a p.m.f.).} \tag{7.6}$$

[1] If a sequence a_n is monotone and bounded, then a_n converges to a limit l. That is, if $a_n \leq a_{n+1}$ or $a_n \geq a_{n+1}$ and $|a_n| \leq M$ for all n and $M > 0$, then $a_n \to l$.

[2] When $0 \leq a_n \leq 1$ in $G(x)$ for all n defined for $x \in (-1, 1)$, we call $G(x)$ the generating function of the sequence $(a_n)_{n\geq 0}$. If a_n forms a probability mass function, then $G(x)$ is called the probability generating function (p.g.f.). The function $G(\theta_k)$ in (7.4) $= \sum_{n=0}^{\infty} p_n\theta_k^n$ is a p.g.f. as $p_0, p_1, p_2, \ldots$ forms a p.m.f. For example, for a Poisson distribution (which has a discrete random variable) with parameter b, the p.g.f. is expressed as

$$P_g(s) = \sum_{n=0}^{\infty} \frac{e^{-b}b^n}{n!} s^n = e^{-b} \sum_{n=0}^{\infty} \frac{(bs)^n}{n!} = e^{-b}e^{bs} = e^{-b(1-s)}.$$

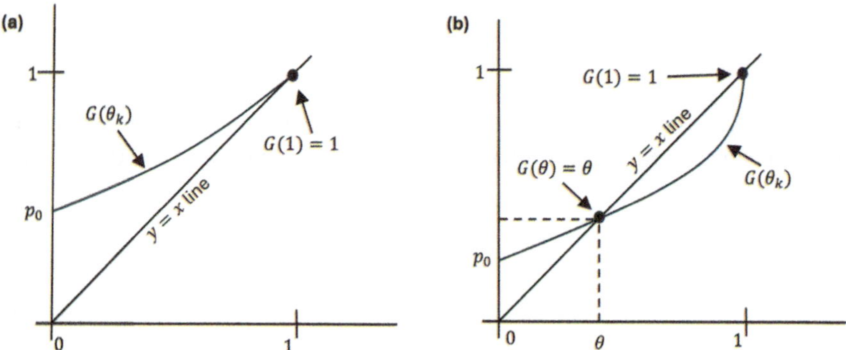

Fig. 7.1 The graph of $G(\theta_k)$ and fixed points: (**a**) The situation $G(1) = 1$ holds, and (**b**) the situation $G(d) = d$ holds for $d \in (0, 1)$ and $G(1) = 1$ holds

Since $G(1) = 1$, we conclude that 1 is a fixed point[3] of G. Also,

$$G'(\theta_k) = p_1 + 2p_2\theta_k + 3p_3\theta_k^2 + \dots. > 0, \tag{7.7}$$

and

$$G''(\theta_k) = 2p_2 + 6p_3\theta_k + 12p_4\theta_k^2 + \dots. > 0. \tag{7.8}$$

These imply G' and G'' are increasing functions because each term in (7.7) and (7.8) are nonnegative. So G is a convex function.[4] Using (7.3), we can write $G(\theta)$ in terms of $\lim \theta_k$ as follows:

$$\begin{aligned} G(\theta) &= G\left(\lim_{k\to\infty} \theta_k\right) \\ &= \lim_{k\to\infty} G(\theta_k) \\ &= \lim_{k\to\infty} \theta_{k+1} \text{ (using (7.5))} \\ &= \theta \text{ (using (7.3))}. \end{aligned} \tag{7.9}$$

Since $G(\theta) = \theta$ by (7.9), we have another fixed point of G that lies in the interval $(0, 1)$ besides earlier fixed point $1 \in [0, 1]$ that we saw above. These two fixed points on the graph of G are shown in Fig. 7.1.

[3] If $f(x) = x$ for any x defined on the domain of f, we say x is a fixed point of $f(x)$.

[4] A function $f : (a, b) \to \mathbb{R}$ is convex, if $f(\alpha x+(1-\alpha)y) \le \alpha f(x)+(1-\alpha)f(y)$ for $x, y \in (a, b)$ and $\alpha \in (0, 1)$. If f is differentiable, and the derivative of f, i.e., f' is increasing, then f is convex. Also, f is convex if the derivative f' is a monotonically increasing function. All convex functions defined on an open interval are continuous on that interval.

Observation III: Let us consider $G'(\theta_k)$ in (7.7) and express $G'(\theta_k)$ for $\theta_k = 1$ as follows:

$$G'(1) = p_1 + 2p_2 + 3p_3 + \ldots = \Sigma_{n=0}^{\infty} np_n. \tag{7.10}$$

7.2 Extinction Probability Theorem

The function $G'(1) = \Sigma_{n=0}^{\infty} np_n$ represents the expected value or the mean value of a discrete random variable whose probabilities are $p_0, p_1, p_2, \ldots$. Therefore, $G'(1)$ is the average number of male offspring born to a couple in a given generation, and this quantity is often denoted by m, i.e., $m = G'(1)$. We state a classical theorem on extinction probabilities in discrete time branching process described above by introducing conditions on $G'(1)$ or m.

Theorem 7.1 (Ultimate Extinction Probability Theorem) *If $m \leq 1$, then $G(1) = 1$, and if $m > 1$, then there exists a unique θ for $\theta \in (0, 1)$ such that $G(\theta) = \theta$, and $G(1) = 1$.*

Proof Let $m \leq 1$. This implies

$$G'(x) < 1 \text{ for } x \in [0,1). \tag{7.11}$$

Integrating (7.11) from y to 1, we get

$$\begin{aligned}
&\int_y^1 G'(x)dx < \int_y^1 dx, \\
&\implies G(1) - G(y) < 1 - y \\
&\qquad 1 - G(y) < 1 - y \text{ (since } G(1) = 1) \\
&\qquad \implies G(y) > y \text{ for } y \in [0, 1)).
\end{aligned} \tag{7.12}$$

Since $G(y) > y$ for all $y \in [0, 1)$ in (7.12) and the graph G is above $y = x$ line for $m \leq 1$ (similar to Fig. 7.1a). Moreover, we know that $G(1) = 1$. This implies 1 is the only fixed point of $G(x)$ for $x \in [0, 1]$.

Let $m > 1$. This implies $G'(1) > 1$. We will see that $G(x)$ has two fixed points in $[0, 1]$, one is value 1 and other is θ in $(0, 1)$, i.e., $\theta < 1$. We saw earlier that both $G(x)$ and $G'(x)$ are strictly increasing functions and continuous on the interval $[0, 1]$. We have for $0 < x_1 < y < 1$,

$$1 < G'(x) < G'(1) = m. \tag{7.13}$$

Integrating (7.13) from y to 1, we get

$$G(y) < y \text{ for } y \in (x_1, 1)). \tag{7.14}$$

Since $G(y) < y$ for all $y \in (x_1, 1)$ in (7.14), the graph G is below the $y = x$ line for $m > 1$ (similar to Fig. 7.1b). Let $x_2 \in (x_1, 1)$.

Consider a new function $H(x) = G(x) - x$. Note that

$$H(0) = G(0) - 0 = p_0 > 0, \tag{7.15}$$

$$H(1) = G(1) - 1 = 1 - 1 = 0, \tag{7.16}$$

$$H(x_2) = G(x_2) - x_2 < 0, \text{ (by (7.14), } G(y) < y \text{ for } y \in (x_1, 1)). \tag{7.17}$$

By (7.15) and (7.17), $H(0) > 0$ and $H(x_2) < 0$. Using intermediate value property[5] of real analysis, there exists a value, say θ in $(x_2, 1)$, for which $H(\theta) = 0$. This implies $G(\theta) = \theta$, that is, θ is a fixed point of G for θ in $(x_2, 1)$.

Next, we will check whether θ is the unique fixed point in the interval $(0, 1)$. Suppose there exists another fixed point, say α in $(0, 1)$, such that $\alpha \neq \theta$. Then $G(\alpha) = \alpha$. Then, α must satisfy either $0 < \alpha < \theta$ or $\theta < \alpha < 1$. Let us evaluate the function H at θ, α, and 1 on the interval $[0, 1]$. These functional values are

$$\begin{aligned} H(\theta) &= G(\theta) - \theta = 0 \text{ (because } \theta \text{ is a fixed point)} \\ H(\alpha) &= G(\alpha) - \alpha = 0 \text{ (because } \alpha \text{ assumed to be another fixed point)} \\ H(1) &= G(1) - 1 = 0. \end{aligned} \tag{7.18}$$

By the mean value theorem (Rolle's),[6] there exists a θ_a in $(0, \alpha)$ and θ_b in $(\alpha, 1)$, such that $H'(\theta_a) = 0$, and $H'(\theta_b) = 0$. This implies

$$G'(\theta_a) = 1, \text{ and } G'(\theta_b) = 1 \text{ for some } \theta_a, \theta_b < 1. \tag{7.19}$$

The conclusion in (7.19) is a contradiction because G' is strictly an increasing function in the interval $[0, 1]$ and only $G'(1) = 1$. For no other values in $[0, 1]$ the function G' can not be equal to 1. Therefore, α cannot be another fixed point in $(0, 1)$. □

A few examples are demonstrated below for the computations of the ultimate extinction probability using the discrete-time branching process framework explained above.

---

[5] Intermediate value property: We say a function f has the intermediate value property on an interval $[a, b]$ in $\mathbb{R}$ if for all x, y in $[a, b]$ and $x < y$ and $f(x) < I < f(y)$, then there exists a point c in (x, y) for which $f(c) = I$.

[6] Rolle's mean value theorem: Let $f : [a, b] \to \mathbb{R}$ continuous on $[a, b]$ and f is differentiable on (a, b), and if $f(a) = f(b)$, then there exists a c in (a, b) such that $f'(c) = 0$.

Example 7.2 Let $p_0 = \frac{1}{3}, p_1 = \frac{2}{9}, p_2 = \frac{4}{9}, p_n = 0$ for $n \geq 3$ be the male offspring probabilities of a branching process. Let us compute the ultimate extinction probability for the family surname for this population. For this, we need to make $G(x) - x = 0$ and solve for x. Compute $G'(1)$ to obtain the average male offspring born to a couple in that generation of population. We have

$$G(x) = p_0 + p_1 x + p_2 x^2$$
$$= \frac{1}{3} + \frac{2}{9}x + \frac{4}{9}x^2.$$

$$G(x) - x = 0 \implies \frac{1}{3} + \frac{2}{9}x + \frac{4}{9}x^2 - x = 0,$$
$$\implies 4x^2 - 7x + 3 = 0,$$
$$=(4x - 3)(x - 1) = 0, \tag{7.20}$$
$$x = \frac{3}{4} \text{ or } x = 1.$$

We conclude that the probability of ultimate extinction is $3/4$.

The mean number of male offspring is computed using (7.10) as

$$m = G'(1) = p_1 + 2p_2 = \frac{2}{9} + 2.\frac{4}{9} = \frac{10}{9} \text{ (i.e., } m > 1). \tag{7.21}$$

Since $m > 1$ for the given offspring probabilities (as we saw in (7.21)), the two fixed points for G by Theorem 7.1 are $x = 3/4$ and $x = 1$, and

$$G\left(\frac{3}{4}\right) = \frac{1}{3} + \frac{2}{9}.\frac{3}{4} + \frac{4}{9}.\left(\frac{3}{4}\right)^2 = \frac{3}{4} \text{ (a fixed point} < 1),$$
$$G(1) = \frac{1}{3} + \frac{2}{9}.1 + \frac{4}{9}.1^2 = 1 \text{ (1 is a fixed point).}$$

Example 7.3 Suppose that $p_0 = \frac{7}{25}, p_1 = \frac{8}{25}, p_2 = \frac{2}{5}, p_n = 0$ for $n \geq 3$ are the male offspring probabilities of a branching process. Let us compute the extinction probabilities, the average male offspring in the generation, and fixed points of G as follows:

$$G(x) = \frac{7}{25} + \frac{8}{25}x + \frac{2}{5}x^2.$$

$$G(x) - x = 0 \implies$$

$$\frac{7}{25} + \frac{8}{25}x + \frac{2}{5}x^2 - x = 0,$$

$$= (10x - 7)(x - 1) = 0,$$

$$x = \frac{7}{10} \text{ or } x = 1. \tag{7.22}$$

The mean number of male offspring is computed using (7.10) as

$$m = G'(1) = p_1 + 2p_2 = \frac{8}{25} + 2.\frac{10}{25} = \frac{28}{25} > 1,$$

and the two fixed points of G (because $m > 1$ and due to Theorem 7.1) for the given probability distribution are

$$G\left(\frac{7}{10}\right) = \frac{7}{25} + \frac{8}{25}.\frac{7}{10} + \frac{2}{5}\left(\frac{7}{10}\right)^2 = \frac{7}{10} \text{ (a fixed point in the interval (0, 1))},$$

$$G(1) = \frac{7}{25} + \frac{8}{25}.1 + \frac{2}{5}1^2 = 1 \text{ (the second fixed point).}$$

Examples 7.2 and 7.3 help to understand the computations of branching process extinction probabilities and to demonstrate the steps involved in such computation. Refer to [1–3] for historical remarks on the branching process, its central role in the field of probability, and applications in population biology and other natural sciences. There were also corrections on historical credits on the branching process's criticality theorem [4, 5]. For further references on branching process applications in demography, refer to [6]. The concepts of the branching process are interesting to mathematical analysts and probabilists working in applications of extinction probabilities. For further reading on uniform convergence of functions, refer to [7, Ch. 9], [2, Ch. 6], [8, Ch. 9], and for other real-world applications of branching process beyond demography, refer to [3, 4, 9].

7.3 Exercises

Exercise 7.4 Let $p_0 = \frac{7}{32}, p_1 = \frac{17}{32}, p_2 = \frac{1}{4}, p_3 = 0, p_4 = 0, \ldots$ be the male offspring probabilities of a branching process. Obtain the mean number of male offspring. Find the ultimate probability of extinction.

Exercise 7.5 Let $p_0 = \frac{5}{31}, p_1 = \frac{15}{31}, p_2 = \frac{11}{31}, p_n = 0$ for $n \geq 3$ be the male offspring probabilities of a branching process. Obtain the mean number of male offspring. Find the ultimate probability of extinction.

Exercise 7.6 Let $G(t) = p_0 + p_1 t + p_2 t^2 + \ldots.$ be the $p.g.f.$ of the offspring distribution satisfying the following properties: (i) $G(0) = p_0 > 0$, $G(1) = 1$, and $p_0 + p_1 < 1$; (ii) $G(t)$ is continuous on $[0, 1]$; (iii) $G'(t), G''(t), G'''(t), \ldots$ are differentiable for $t \in [0, 1)$; (iv) $G'(t) > 0$ for $t \in (0, 1]$; $and (v)$ $G''(t) > 0$ for $t \in (0, 1)$. Prove that $G'(t) < 1$ for $t \in [0, 1)$ if, and only if, $p_1 + 2p_2 + 3p_3 + \ldots \leq 1$.

References

1. Karlin, S.: A First Course in Stochastic Processes. Academic Press, New York, London (1996)
2. Abbott, S.: Understanding analysis. Undergraduate Texts in Mathematics, xii+257 pp. Springer, New York (2001)
3. González, M., Puerto, I.M., Martínez, R., Molina, M., Mota, M., Ramos, A. (Eds.): Workshop on Branching Processes and Their Applications, Vol. 197. Springer Science & Business Media (2010)
4. Srinivasan, K., Reddy, P.H., Raju, K.M.: From one generation to the next: Changes in fertility, family size preferences, and family planning in an Indian state between 1951 and 1975. Stud. Family Plann. **9**(10/11), 258–271 (1978)
5. Kendall, D.G.: The genealogy of genealogy branching processes before (and after) 1873. Bull. Lond. Math. Soc. **7**(3), 225–253 (1975)
6. Chu, C.Y.C.: Demographic Models and Branching Processes. Population Dynamics: A New Economic Approach (New York, 1998; online edn.) Oxford Academic (2020)
7. Apostol, T.M.: Mathematical Analysis, 2/e. Addisson-Wesley, New York and Narosa Publishers (India), New Delhi (1973)
8. Krantz, S.G.: Real analysis and foundations. Fifth edition [of MR1210958]. Textbooks in Mathematics, xv+484 pp. CRC Press, Boca Raton, FL (2022). ISBN:978-1-032-10272-6; 978-1-032-12026-3; 978-1-003-22268-2
9. Rao, A.S.R.S.: Probabilities of therapeutic extinction of HIV. Appl. Math. Lett. **19**(1), 80–86 (2006)

References

Chapter 8
Multiple Decrement Table

The life table construction that we learned in Chap. 3 was a single decrement table. The concept of a life table or a single decrement table can be extended (as we mentioned in Chap. 3) to various other situations other than death distribution data over the age a. Requirement for a multiple decrement table arises in several contexts, for example, when we can distribute total deaths in a single decrement table into $n-$independent causes of deaths ($n = 2, 3, ..$), when we can distribute total unhealthy or sick individuals in a single decrement morbidity table into $n-$independent morbid causes ($n = 2, 3, ..$), etc.

8.1 Decrement Table Through k Causes

We construct a multiple decrement table if we can distribute total deaths d_a during the age interval $[a, a + 1)$ of a single decrement table into $d_a^{(1)}, d_a^{(2)}, \ldots, d_a^{(n)}$ independent causes of deaths or if we can distribute the total force of mortality μ_a into n independent forces of mortalities $\mu_a^{(1)}, \mu_a^{(2)}, \ldots, \mu_a^{(n)}$. That is,

$$d_a = d_a^{(1)} + d_a^{(2)} + \ldots + d_a^{(n)}, \tag{8.1}$$

or

$$\mu_a = \mu_a^{(1)} + \mu_a^{(2)} + \ldots + \mu_a^{(n)}, \tag{8.2}$$

where $d_a^{(k)}$ is the number of deaths during the age interval $[a, a + 1)$ due to the cause k, and $\mu_a^{(k)}$ is the force of mortality during the age interval $[a, a + 1)$ due

A. S. R. Srinivasa Rao, *Mathematical Demography: Theory and Modeling*, Mathematical Marvels: Texts and Monographs in the Spirit of CR Rao,
https://doi.org/10.1007/978-981-95-6129-2_8

to the cause k for $k = 1, 2, \ldots, n$. Using (8.1) and (8.2), we can construct other life table functions similar to the functional relations we saw in Chap. 3. The k independent death probabilities $q_a^{(k)}$ are

$$q_a^{(k)} = \frac{d_a^{(k)}}{l_a} \text{ for } k = 1, 2, \ldots, n, \tag{8.3}$$

and the corresponding death rates are

$$m_a^{(k)} = \frac{2q_a^{(k)}}{2 - q_a^{(k)}} \text{ for } k = 1, 2, \ldots, n. \tag{8.4}$$

Using (3.6), we write

$$d_a = \int_0^1 l_{a+t}\mu_{a+t}dt. \tag{8.5}$$

Dividing both sides of (8.5) by l_a, we get

$$\frac{d_a}{l_a} = \int_0^1 \frac{l_{a+t}}{l_a}\mu_{a+t}dt,$$

which implies

$$q_a = \int_0^1 {}_tp_a\mu_{a+t}dt, \tag{8.6}$$

because from Chap. 3 we saw that $q_a l_a = d_a$, and $\frac{l_{a+t}}{l_a} = {}_tp_a$. The continuous form for $q_a^{(k)}$ is similar to the formula in (8.6), and it can be written using the force of mortality $\mu_a^{(k)}$ as follows:

$$q_a^{(k)} = \int_0^1 {}_tp_a\mu_{a+t}^{(k)}dt. \tag{8.7}$$

By (3.16), we have

$$p_a = \exp\left(-\int_0^a \mu_s ds\right). \tag{8.8}$$

Substituting the expression of $\mu_s = \mu_s^{(1)} + \mu_s^{(2)} + \ldots + \mu_s^{(n)}$ in (8.8), we get

$$p_a = \exp\left\{-\int_0^a \left(\mu_s^{(1)} + \mu_s^{(2)} + \ldots + \mu_s^{(n)}\right) ds\right\},$$

$$= \exp\left(-\int_0^a \mu_s^{(1)} ds\right) \exp\left(-\int_0^a \mu_s^{(2)} ds\right) \ldots \exp\left(-\int_0^a \mu_s^{(n)} ds\right),$$

$$= p_a^{(1)} . p_a^{(2)} \ldots . p_a^{(n)} = \prod_{k=1}^{n} p_a^{(k)}. \tag{8.9}$$

Example 8.1 Suppose d_0 in Table 3.1 is distributed into three causes $d_0^{(1)}$, $d_0^{(2)}$, and $d_0^{(3)}$, where $d_0^{(1)} = 2$, $d_0^{(2)} = 1$, and $d_0^{(3)} = 1$, respectively. Then

$$q_0^{(1)} = \frac{d_0^{(1)}}{l_0} = \frac{2}{25} = 0.08,$$

$$q_0^{(2)} = \frac{d_0^{(2)}}{l_0} = \frac{1}{25} = 0.04,$$

$$q_0^{(3)} = \frac{d_0^{(3)}}{l_0} = \frac{1}{25} = 0.04.$$

We can obtain other values of the multiple decrement table once we know the distribution of the causes of deaths at all the ages $d_1^{(k)}, d_2^{(k)}, \ldots$ for $k = 1, 2, 3$ in Table 3.1.

Let $l_a^{(k)}$ represent the number of individuals living at age a who will be eventually removed from the life table due to cause k; then

$$l_a \left(= \sum_{k=1}^{n} l_a^{(k)}\right) = l_0 \exp\left(-\int_0^a \mu_s ds\right),$$

$$\Longrightarrow {}_n p_a = \exp\left(-\int_a^{a+n} \mu_s ds\right) = \exp\left(-\int_0^n \mu_{a+t} dt\right), \tag{8.10}$$

where μ_s in (8.10) is the total force of mortality due to all causes defined in (8.2). Here $l_a^{(k)}$ can also be represented as $l_a^{(k)} = \sum_{b=a}^{\omega} d_b^{(k)}$ for $k = 1, 2, \ldots, n$.

Now, let us introduce a slightly different notation for understanding the cause of death data from a multiple decrement table. Let $l_a^{*(k)}$ represent the l_a column of an associated single-decrement life table corresponding to a cause k of decrement (or death corresponding to cause k) of a multiple decrement table. We define $l_a^{*(k)}$ as

$$l_a^{*(k)} = l_0^{*(k)} \exp\left(-\int_0^a \mu_s^{(k)} ds\right), \tag{8.11}$$

where $l_0^{*(k)}$ is the radix of the corresponding single-decrement table for cause k. From (8.11) we have

$$p_a^{*(k)} = \exp\left(-\int_0^a \mu_s^{(k)} ds\right),$$
$$q_a^{*(k)} = 1 - \exp\left(-\int_0^a \mu_s^{(k)} ds\right). \tag{8.12}$$

Example 8.2 Under the uniform death distribution (UDD), show that $q_a^{*(k)} = 1 - (p_a)^{\frac{d_a^{(k)}}{d_a}}$, where d_a is the total deaths at age a due to all causes defined in (8.1).

Solution We note that

$$q_a^{*(k)} = 1 - \exp\left(-\int_0^1 \mu_{a+t}^{(k)} dt\right)$$
$$\text{because}\left(q_a^{*(k)} = 1 - \exp\left(-\int_0^a \mu_s^{(k)} ds\right) \text{ from (8.12)}\right)$$
$$= 1 - \exp\left(-\int_0^1 \frac{d_a^{(k)}}{l_a - td_a} dt\right)$$
$$= 1 - \exp\left\{-\frac{d_a^{(k)}}{d_a} \log(l_a - td_a)\right\}\Bigg|_0^1 = 1 - \exp\left(-\frac{d_a^{(k)}}{d_a} \log p_a\right)$$
$$= 1 - (p_a)^{\frac{d_a^{(k)}}{d_a}}.$$

Under the uniform death distribution, we write l_{a+t} as

$$l_{a+t} = l_a - t(l_a - l_{a+1}),\ (0 \le t \le 1), \tag{8.13}$$

where $d_a = l_a - l_{a+1}$. Under the assumption of UDD the deaths in a year are spread evenly throughout the year. We can associate $m_a^{(k)}$ the central death rate at age a due to cause k with $d_a^{(k)}$ as

$$d_a^{(k)} = d_a \frac{m_a^{(k)}}{m_a}, \tag{8.14}$$

where m_a is the central death rate at age a due to all causes. We can compute $m_a^{(k)}$ using

$$m_a^{(k)} = d_a^{(k)} \frac{m_a}{d_a} \left(\text{here } d_a^{(k)} = \int_0^1 l_{a+t}\mu_{a+t}^{(k)} dt\right). \tag{8.15}$$

Similarly,

$$q_a^{*(k)} = \frac{d_a^{(k)}}{L_a + \frac{1}{2}d_a^{(k)}} = \frac{\frac{d_a^{(k)}}{L_a}}{1 + \frac{1}{2}\frac{d_a^{(k)}}{L_a}}$$

$$= \frac{m_a^{(k)}}{1 + \frac{1}{2}m_a^{(k)}} \left(\text{since } \frac{d_a^{(k)}}{L_a} = m_a^{(k)} \right). \tag{8.16}$$

This implies

$$m_a^{(k)} = \frac{2q_a^{*(k)}}{2 - q_a^{*(k)}}. \tag{8.17}$$

8.2 Survival Functions and Hazard Rate

The force of mortality function in demography is known as the hazard rate in survival analysis of data. We have seen the definition of the force of mortality and various associated expressions in Chap. 3. The distribution function $F_a(t)$ of T_a (the remaining lifetime of an individual age a) is represented by

$$F_a(t) = P\left(T_a \leq t\right) = 1 - {}_tp_a = {}_tq_a \; (0 \leq a \leq \omega). \tag{8.18}$$

The survival function of an individual aged a is defined as

$${}_tp_a = 1 - F_a(t) = P\left(T_a > t\right). \tag{8.19}$$

See Fig. 8.1. The survival function (${}_tp_0$) and distribution function ($F_0(t)$) of T_0 (the remaining lifetime of a newborn individual) can be related with $F_a(t)$ as follows:

$${}_tq_a = F_a(t) = P\left(T_a \leq t\right) = P\left(T_0 \leq a + t / T_0 > a\right) \tag{8.20}$$

$$= \frac{P\left\{(T_0 \leq a + t) \cap (T_0 > a)\right\}}{P\left(T_0 > a\right)}. \tag{8.21}$$

Using the relationship between ${}_tq_a$ and μ_a studied in Chap. 3, we can represent

$${}_{\Delta t}q_a = \Delta t \mu_a \text{ (when } \Delta t \text{ is very small).} \tag{8.22}$$

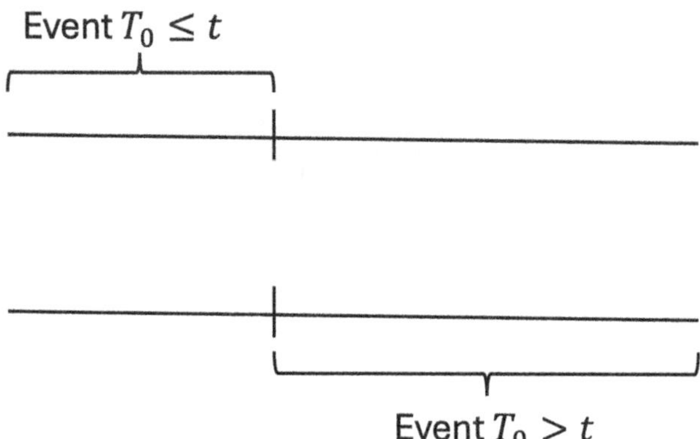

Fig. 8.1 Remaining lifetime of newborn individuals

That is, the probability of death within a small interval is proportional to the interval length. Using (8.21) and expanding the definition of μ_a to μ_{a+t}, we can write

$$\begin{aligned}\mu_{a+t} &= \lim_{\Delta t \to 0^+} \frac{{}_{\Delta t}q_{a+t}}{\Delta t} = \lim_{\Delta t \to 0^+} \frac{1}{\Delta t} P\left(T_0 \le a+t+\Delta t / T_0 > a+t\right) \\ &= \lim_{\Delta t \to 0^+} \frac{1}{\Delta t} P\left(T_a \le t+\Delta t / T_a > t\right). \end{aligned} \tag{8.23}$$

Example 8.3 Show that $\frac{d}{dt} F_a(t) = {}_t p_a \mu_{a+t} \ (0 \le t < \omega - a)$.

Proof Let us write $\frac{d}{dt} F_a(t)$ using the definition of derivative as follows:

$$\begin{aligned}\frac{d}{dt} F_a(t) &= \frac{d}{dt} P\left(T_a \le t\right) \\ &= \lim_{\Delta t \to 0^+} \frac{1}{\Delta t} \left\{P\left(T_a \le t+\Delta t\right) - P\left(T_a \le t\right)\right\}. \end{aligned} \tag{8.24}$$

Using the conditional probability of (8.20) in (8.24), we get

$$\begin{aligned}\frac{d}{dt} F_a(t) &= \lim_{\Delta t \to 0^+} \frac{1}{\Delta t} \left\{P\left(T_0 \le a+t+\Delta t / T_0 > a\right) - P\left(T_0 \le a+t / T_0 > a\right)\right\} \\ &= \lim_{\Delta t \to 0^+} \frac{1}{\Delta t} \left[\frac{P\left\{\left(T_0 \le a+t+\Delta t\right) \cap \left(T_0 > a\right)\right\}}{P\left(T_0 > a\right)} \right. \\ &\qquad \left. - \frac{P\left\{\left(T_0 \le a+t\right) \cap \left(T_0 > a\right)\right\}}{P\left(T_0 > a\right)}\right] \\ &= \lim_{\Delta t \to 0^+} \frac{\left\{P\left(T_0 \le a+t+\Delta t\right) - P\left(T_0 \le a+t\right)\right\}}{\Delta t P\left(T_0 > a\right)}. \end{aligned} \tag{8.25}$$

Multiplying and dividing (8.25) by $P(T_0 > a+t)$, we get

$$\frac{d}{dt}F_a(t) = \frac{P(T_0 > a+t)}{P(T_0 > a)} \lim_{\Delta t \to 0^+} \frac{\{P(T_0 \le a+t+\Delta t) - P(T_0 \le a+t)\}}{\Delta t P(T_0 > a+t)}$$
$$= {}_t p_a \mu_{a+t} \; (0 \le t < \omega - a).$$

□

Using the relation in Example 8.3, we can compute the expected future lifetime of T_a denoted by $E[T_a]$ as follows:

$$\begin{aligned} E[T_a] &= \int_0^{\omega-a} t \frac{d}{dt} F_a(t) dt \\ &= \int_0^{\omega-a} t {}_t p_a \mu_{a+t} dt \\ &= \int_0^{\omega-a} t \left(-\frac{\partial}{\partial t} {}_t p_a \right) dt. \end{aligned} \tag{8.26}$$

Integrating by parts with respect to t, we get

$$= -t_t p_a|_0^{\omega-a} + \int_0^{\omega-a} {}_t p_a dt = \int_0^{\omega-a} {}_t p_a dt. \tag{8.27}$$

In (8.26), we used the fact that $\frac{\partial}{\partial t} {}_t p_a = \frac{\partial}{\partial t}(1 - F_a(t)) = -\frac{\partial}{\partial t} F_a(t) = \frac{\partial}{\partial t} {}_t p_a \mu_{a+t}$. From Exercise 8.9, we can write ${}_1 p_a = \exp\left(-\int_0^1 \mu_{a+s} ds\right)$. This implies $-\int_0^1 \mu_{a+s} ds = \log_1 p_a$. We approximate the mean value of the force of mortality between a and $a+1$ as $\mu_{a+\frac{1}{2}} \approx -\log_1 p_a$.

Efforts have been made to model the force of mortality as a function of multiple variables to account for various causes of decline [1]. These multivariate force of mortality ideas are not discussed in this book. Multiple decrement tables are also frequently used in the insurance industry to analyze the impacts of various reasons for failure of automobiles and other subjects [1–6].

The procedure described above can be useful for estimating the remaining life expectancy associated with each cause of death or each cause of decrement. However, if the causes of death or decrements are not independent and if there are confounding factors affecting these causes, the multiple decrement framework outlined earlier may not provide accurate estimates.

8.3 Exercises

Exercise 8.4 In a multiple decrement table, the force of mortality at age a for cause k is given by a function $\mu_a^{(k)} = \frac{1}{ka}$ for $k = 1, 2, 3, 4$. Find the probability that an individual aged 10 will live until age 85.

Exercise 8.5 A multiple decrement table is constructed based on two attributes in the population, namely, $k = 1$ (smokers with lung diseases) and $k = 2$ (chewing tobacco-related mouth cancer) with a force of mortality function defined by $\mu_a^{(k)} = \frac{k}{100-a}$ for $k = 1, 2$. Assuming causes of death are independent of each other, find the probability of death for the individuals aged 85 to die before they reach the age of 90.

Exercise 8.6 Let ${}_t p_a$ be the survival probability function at age a for all causes of death in a multiple decrement table. Show that (a) ${}_n q_a^{(k)} = \int_0^n {}_t p_a \mu_{a+t}^{(k)} dt$, and $\mu_a^{(k)} = -\frac{1}{l_a} \frac{dl_a^{(k)}}{da}$.

Exercise 8.7 Suppose we are interested in constructing a single-decrement table from multiple decrement functions. Let p_a be the survival probability due to all causes, $d_a^{(k)}$ be the number of deaths at age a due to cause k, d_a be the number of deaths due to all causes, $q_a^{(k)}$ be the probability of dying at age a due to cause k, and $q_a^{*(k)}$ be the associated single-decrement function representing probability of dying; then by ignoring $q_a^3, q_a^4, \ldots$, show that

$$q_a^{*(k)} = \frac{d_a^{(k)}}{d_a} q_a - \frac{1}{2} \frac{d_a^{(k)}}{d_a} \left(\frac{d_a^{(k)}}{d_a} - 1 \right) (q_a)^2 .$$

Exercise 8.8 Let T_a represent a random variable describing a complete future lifetime. Evaluate var$[T_a]$ in terms of $E[T_a]$ and μ_{a+t}. (here $E[T_a]$ is the expectation of T_a.)

Exercise 8.9 Show that ${}_t p_a = \exp\left(- \int_0^t \mu_{a+s} ds \right)$.

Hint. $\frac{\partial}{\partial t} {}_t p_a = \frac{\partial}{\partial t}(1 - {}_t q_a) = -{}_t p_a \mu_{a+t}$.

Exercise 8.10 Given a multiple decrement table with k causes, show that

$$d_a^{(k)} = \frac{q_a^{(k)}}{q_1^{(k)} + q_a^{(k)}} d_a, \text{ for } k = 1, 2,$$

where d_a is the quantity representing the total deaths due to all causes.

Exercise 8.11 Given the survival function $s(a) = \frac{1}{200}(200 - 2a - 0.5a^2)$, find the probability of (i) survival up to age $80 and (ii)$ a person aged 75 would live until age 95.

References

1. Mitra, S., Singh, P., Rao, A.S.R.S.: Multivariate Force of Mortality. Demography India, Vol. 44(1&2), pp. 01–16 (2015). ISSN: 0970-454X. (Also available at arXiv preprint arXiv:1111.5213) (2019)
2. Yusuf, F., Martins, J.M., Swanson, D.A., Yusuf, F., Martins, J.M., Swanson, D.A.: Multiple decrement life tables. Methods Demographic Anal., 215–229 (2014)
3. Deshmukh, S.: Multiple Decrement Models. In: Multiple Decrement Models in Insurance. Springer, India (2012). https://doi.org/10.1007/978-81-322-0659-0_1
4. Arnold, B.C., Brockett, P.L.: Identifiability for dependent multiple decrement/competing risk models. Scand. Actuarial J. **1983**(2), 117–127 (1983)
5. Hanagal, D.D.: Modeling Survival Data Using Frailty Models, xx+314 pp. CRC Press, Boca Raton, FL (2011)
6. Dodd, E., Forster, J.J., Bijak, J., Smith, P.W.F.: Stochastic modelling and projection of mortality improvements using a hybrid parametric/semi-parametric age-period-cohort model. Scand. Actuar. J. **2021**(2), 134–155 (2021)

References

Correction to: Mathematical Demography: Theory and Modeling

Correction to: A. S. R. Srinivasa Rao, *Mathematical Demography: Theory and Modeling*, Mathematical Marvels: Texts and Monographs in the Spirit of CR Rao, https://doi.org/10.1007/978-981-95-6129-2

After initial publication of the book, various errors were identified that needed correction. All corrections listed below have been updated within the current version.

Corrections:

- **Frontmatter:**
  Page ix, Line 3: Kindly replace "Yeuhua" with "Yuehua."
  Line 6: Kindly add the missing PE and PS names.
- **Chapter 1:**
  Page 5, Line 3 from the top and below table 1.1: Replace "When" with "if" in both places.
  Page 6, Line 1 from the bottom (Equation 1.17): Replace "=" with " $\approx$".
  Page 7, Line 3 from the bottom: Delete the repeated word "generation".
  Page 10, Line 3 from the bottom: Add a dot after Equation 1.37.
- **Chapter 2:**
  Page 26, Line 7 from the top (Equation 2.28): Replace "+" with "-".
- **Chapter 3:**
  Page 46, Line 1 from the top below table 3.2: Remove the space between "d_a" and ",".

The updated version of this book can be found at
https://doi.org/10.1007/978-981-95-6129-2

A. S. R. Srinivasa Rao, *Mathematical Demography: Theory and Modeling*,
Mathematical Marvels: Texts and Monographs in the Spirit of CR Rao,
https://doi.org/10.1007/978-981-95-6129-2_9

- **Chapter 4:**
  Page 61, Line 4 from the top: Replace "X" with "1/x".
- **Chapter 5:**
  Page 87, line 13 from the top: Remove the space between "SPI" and ",".
  Page 90, Figure 5.5: Move the value "0.18" to the right.
  Page 99, Line 14 from the bottom: Replace "Leigh" with "Length".
- **Chapter 6:**
  Page 117, Line 4 from the bottom: Replace "From Footnote 2, we" with "We".
- **Chapter 7:**
  Page 126, Line 11 from the bottom: Replace "spring" with "offspring".
  Page 131, Line 6 from the bottom: Add a comma.
- **Backmatter:**
  Appendix - Page 151, Line 3 from the bottom: Insert "d_{10}".

Appendix A
Metric Space

A sequence (a_n) (for $a_n \in \mathbb{R}$) converges to a real number a if, for every positive number ϵ or (for every $\epsilon > 0$), there exists an $N \in \mathbb{N}$ such that whenever $n \geq N$ it follows that $|a_n - a| < \epsilon$. (Here $\mathbb{R}$ is the set of real numbers and $\mathbb{N}$ is the set of natural numbers.) Convergence of a sequence of numbers a_n to a limit a means, given an ϵ, the terms of the sequence a_n after a stage N will be around the point a. A sequence could be denoted by $(a_n)_{n\geq 0}$ if the terms of a sequence start from a_0 or it could be denoted by $(a_n)_{n\geq 1}$ if the terms of a sequence start from a_1. More on convergence of a sequence can be seen in Appendix E.

Definition (Metric Space) Let S be a set. A function $d : S \times S \to \mathbb{R}$ is a metric on S and (S, d) is called a metric space if for all $x, y \in S (i) d(x, y) \geq 0$ with $d(x, y) = 0$ if, and only if, $x = y$, (ii) $d(x, y) = d(y, x)$, and (iii) for all $z \in S, d(x, y) \leq d(x, z) + d(z, y)$.

Here $d(.,.)$ is called the metric of the space (S, d) and d is a nonnegative number. The number d measures the distance between two points x and y for all x, y in (S, d). The property $d(x, y) \leq d(x, z) + d(z, y)$ is often called the triangle inequality. A sequence $a_n \subseteq S$ (for $a_n \in \mathbb{R}$) converges to an element $a \in S$ if for all $\epsilon > 0$ there exists an $N \in \mathbb{N}$ such that $d(a_n, a) < \epsilon$ whenever $n \geq N$. We list below two important properties related to metric spaces and convergence of a sequence (a_n):

(i). A sequence (a_n) in (S, d) converges to l if, and only if, every subsequence of (a_n), i.e., $(a_{k_n})_{n\geq 1}$ converges to l.

(ii). Suppose $(a_n) \to l$ in (S, d). Then for every $\epsilon > 0$, there exists an integer N, such that $d(a_n, a_m) < \epsilon$ whenever $n \geq N$ and $m \geq N$.

Let $\mathbb{R}^n$ be the Euclidean space (where each element is an $n-$tuple). We will come across $\mathbb{R}^n$ in Appendix H in our study of Jacobian matrix and eigenvalues. A point l in $\mathbb{R}^n$ is represented by $(l_1, l_2, \ldots, l_n)$. Let f_1 and f_2 be in $\mathbb{R}^n$ and f_1 and f_2 be

A. S. R. Srinivasa Rao, *Mathematical Demography: Theory and Modeling*, Mathematical Marvels: Texts and Monographs in the Spirit of CR Rao, https://doi.org/10.1007/978-981-95-6129-2

continuous. Let a point l be in (S, d). Then $f_1 + f_2$ is continuous at l, and $f_1 f_2$ is continuous at l.

For further reading on metric spaces, see [1, 2, 8].

Definition A.1 (Probability Space) The triplet $(\Omega, \mathcal{F}, P)$ is called the probability space, where Ω is the sample space (i.e., set of all elementary events or all possible outcomes in an experiment), $\mathcal{F}$ is the *Borel field or a* $\sigma-$ *field* of events in the sample space (Ω), and the probability function P is defined on $\mathcal{F}$. If for any $A \in \mathcal{F}$, we have $P(A) \geq 0$. Suppose $A_1, A_2, \ldots A_k$ is $k-$ elementary and disjoint events in $\mathcal{F}$ such that $\cup_{i=1}^{k} A_i = \Omega$, then $\Sigma_{i=1}^{k} P(A_i) = 1$ or $P(\Omega) = 1$. When $\cup_{i=1}^{\infty} A_i = \Omega$, and $A_i \neq A_j$ for $i \neq j$, then $P(\cup_{i=1}^{\infty} A_i) = P(\Omega) = 1$.

A random variable, say X, is a real-valued function defined on Ω, and it can be written as $X : \Omega \to \mathbb{R}$. Here X could take discrete values $x_1, x_2, \ldots, x_n$ (finite) or $x_1, x_2, \ldots,$ (countable) or continuous values. For an outcome of an experiment, say $\omega \in \Omega$, the distribution function F is defined as

$$F(x) = P(\omega : X(\omega) < x) = P(X < x).$$

For further reading on probability spaces, see Chapter 2 in [6].

Appendix B
Probability Density Function

Let F be a distribution function on a continuous random variable X defined on a probability space $(\Omega, \mathcal{F}, P)$. The probability space $(\Omega, \mathcal{F}, P)$ and random variables are explained in Appendix A. If F is represented by

$$F(x) = P(\omega : X(\omega) < x) = \int_{-\infty}^{x} f(x)dx, \tag{B.1}$$

we call $f(x)$ for $x \in X$ a probability density function (p.d.f.). The quantity $f(x)dx$ is roughly interpreted as the probability that the random variable X belongs to the interval $(x, x+dx)$. Instead of a continuous, if X is a discrete random variable (i.e., X is assuming discrete values, say $x_1, x_2, \ldots, x_n$, or assuming countable discrete values), then F becomes a step function. For a discrete X, we call f a probability mass function (p.m.f.). The p.m.f. is usually denoted by p instead of f, and $p(x_i) = P(X = x_i)$ for $i = 1, 2, \ldots$.

If X follows a normal or a Gaussian distribution with mean μ and variance σ^2, then the p.d.f. of X is

$$f(x) = \frac{1}{\sigma\sqrt{2\pi}} e^{-\frac{(x-\mu)^2}{2\sigma^2}},$$

for $-\infty < \mu < \infty, \sigma^2 > 0$, and $-\infty < x < \infty$. The normal distribution random variable is continuous. If X follows a binomial distribution for $x = 1, 2, \ldots, n$, then the probability mass function of X is

$$p(x) = \binom{n}{x} p^x (1-p)^{n-x}$$

for $x = 0, 1, 2, \ldots n$.

For further reading on probability density functions, see Chapter 2 in [6].

A. S. R. Srinivasa Rao, *Mathematical Demography: Theory and Modeling*, Mathematical Marvels: Texts and Monographs in the Spirit of CR Rao,
https://doi.org/10.1007/978-981-95-6129-2

Appendix C
Complex Conjugates

A complex number z can be expressed as $z = (a, b) = a + ib$, where a and b are real numbers (i.e., $a, b \in \mathbb{R}$), and $i = \sqrt{-1}$. The number i corresponds to (0, 1), which can be written as $0 + i1$, and the number 1 corresponds to $(1, 0)$, or $1 + i0$. The term ib represents the product of i and b. The absolute value of z is obtained by $\sqrt{a^2 + b^2}$. The sum of two complex numbers $z_1 = (a_1, b_1)$ and $z_2 = (a_2, b_2)$ is expressed as $z_1 + z_2 = (a_1 + a_2, b_1 + b_2)$.

The multiplication of two complex numbers, $z_1 = a_1 + ib_1$ and $z_2 = a_2 + ib_2$, involves real numbers a_1, a_2, b_1, and b_2 and is defined as follows:

$$z_1 z_2 = a_1 a_2 - b_1 b_2 + i(a_1 b_2 + a_2 b_1). \tag{C.1}$$

The product $z_1 z_2$ is also written as $(a_1 a_2 - b_1 b_2, a_1 b_2 + a_2 b_1)$. The complex conjugate of z is denoted as $\bar{z} = a - ib$. Refer to Fig. C.1. The product of z and its conjugate $\bar{z}$ is $z\bar{z} = |z|^2$, where $|z| = \sqrt{a^2 + b^2}$. Using (C.1) we note that $i^2 = i.i = (0, 1).(0, 1) = 1 + i.0 = 1$, which gives us $i = \sqrt{-1}$. The complex conjugate of i is $-i$, and the complex conjugate of 1 is 1, and the complex conjugate of $2+3i$ is $2-3i$. In fact, the complex conjugate of any real number a is the number self. The exponential of a complex number is represented as $e^z = e^{a+ib} = e^a.e^{ib} = e^a(\cos b + i \sin b)$. Similarly, the exponential of sum of two complex numbers z_1 and z_2 is expressed as

$$e^{z_1+z_2} = e^{x_1+x_2}(\cos(b_1 + b_2) + i \sin(b_1 + b_2)) = e^{x_1+x_2} e^{i(b_1+b_2)} = e^{z_1} e^{z_2}.$$

For $z \neq 0$, there exists a unique θ such that $a = |z| \cos \theta i$ and $b = |z| \sin \theta$ for $-\pi < \theta \leq \pi$. Here θ is called the principal argument and is represented by $\theta = \arg z$. For an integer n, every z can be represented by $z = re^{i\theta}$,where $r = |z|$ and $\theta = \arg z + 2\pi n$. For a nonzero z and for a positive integer n, there are exactly n distinct roots of z such that $z_0^n = z$, $z_1^n = z, \ldots, z_{n-1}^n = z$. We represent these

A. S. R. Srinivasa Rao, *Mathematical Demography: Theory and Modeling*,
Mathematical Marvels: Texts and Monographs in the Spirit of CR Rao,
https://doi.org/10.1007/978-981-95-6129-2

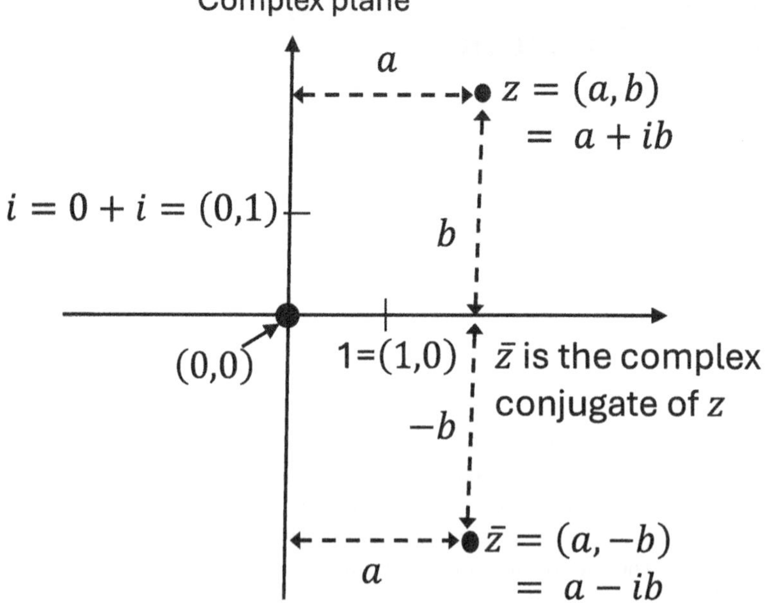

Fig. C.1 Complex conjugates

n roots by $z_k = |z|^{1/n} e^{ih_k}$, where $h_k = \frac{\arg z}{n} + \frac{2\pi k}{n}$ for $k = 0, 1, \ldots, n-1$. Here $z_0^n, z_1^n, \ldots, z_{n-1}^n$ are nth roots of z. Note that $\left(|z|^{1/n} e^{ih_k}\right)^n$

$$= |z|\, e^{inh_k} = |z|\, e^{i(\arg z + 2\pi k)} \text{ for } k = 0, 1, \ldots, n-1.$$

For further reading on complex numbers and their conjugates, see [1, 3–5].

Appendix D
Permutations

Let $A = \{a_1, a_2, a_3\}$, and $B = \{b_1, b_2\}$ be two sets. Suppose we wish to express a new set Ω of distinct elements using the elements both in A and B, and the order of the elements does not matter, i.e., $a_1 b_2$ is treated the same as $b_2 a_1$. The set Ω will be

$$\Omega = \{a_1b_1, a_1b_2, a_2b_1, a_2b_2, a_3b_1, a_3b_2\}. \tag{D.1}$$

In (D.1), the 6 distinct elements (i.e., 3 elements of A times 2 elements of B) of Ω were expressed. Now, let $A = \{a_1, a_2, a_3\}$, $B = \{b_1, b_2\}$, and $C = \{c_1, c_2, c_3, c_4, c_5, c_6, c_7\}$ be three sets, and we aim to express Ω using A, B, and C. The set Ω will be

$$\begin{aligned}\Omega = \{&a_1b_1c_1, a_1b_2c_1, a_2b_1c_1, a_2b_2c_1, a_3b_1c_1, a_3b_2c_1,\\ &a_1b_1c_2, a_1b_2c_2, a_2b_1c_2, a_2b_2c_2, a_3b_1c_2, a_3b_2c_2,\\ &a_1b_1c_3, a_1b_2c_3, a_2b_1c_3, a_2b_2c_3, a_3b_1c_3, a_3b_2c_3,\\ &a_1b_1c_4, a_1b_2c_4, a_2b_1c_4, a_2b_2c_4, a_3b_1c_4, a_3b_2c_4,\\ &a_1b_1c_5, a_1b_2c_5, a_2b_1c_5, a_2b_2c_5, a_3b_1c_5, a_3b_2c_5,\\ &a_1b_1c_6, a_1b_2c_6, a_2b_1c_6, a_2b_2c_6, a_3b_1c_6, a_3b_2c_6,\\ &a_1b_1c_7, a_1b_2c_7, a_2b_1c_7, a_2b_2c_7, a_3b_1c_7, a_3b_2c_7\}.\end{aligned} \tag{D.2}$$

The 42 distinct elements of Ω in (D.2) were listed by joining each of the seven elements of the set C with each of the six elements of Ω in (D.1), i.e., $6 \times 7 = 42$. Let $D = \{d_1, d_2, d_3, d_4, d_5, d_6, d_7, d_8, d_9, d_{10}, d_{11}\}$ be given and we aim to express Ω using A, B, C, and D. As we did in (D.2), the new Ω can be expressed using 42 elements of Ω in (D.2) with 11 distinct elements of D. There will be 462 $(42 \times 11) = (3 \times 2 \times 7 \times 11)$ distinct elements in Ω using the elements of A, B, C, and D.

A. S. R. Srinivasa Rao, *Mathematical Demography: Theory and Modeling*, Mathematical Marvels: Texts and Monographs in the Spirit of CR Rao,
https://doi.org/10.1007/978-981-95-6129-2

Appendix E
Uniform Convergence

Definition E.1 (Uniform Convergence) Let f_n be a sequence of functions defined on a set $S \subseteq \mathbb{R}$. Then, the sequence f_n converges uniformly on S to a limit function f defined on S if, for every $\epsilon > 0$, there exists an $N \in \mathbb{N}$ such that $|f_n(x) - f(x)| < \epsilon$ whenever $n \geq N$ and $x \in S$. Here $\mathbb{R}$ is the set of real numbers, and $\mathbb{N}$ is the set of natural numbers.

That is, when $f_n \to f$ uniformly, after some stage N, the terms of f_n will be within the $\epsilon-$band for a chosen ϵ. If $g_n \to g$ uniformly, then $f_n + g_n$ is also uniformly convergent, but $f_n g_n$ need not be. Next, let us understand uniform convergence of a series.

For a sequence of real or complex numbers $(a_n)_{n \geq 1}$ if we write $s_n = \sum_{n=1}^{\infty} a_n$, then s_n is called an infinite series of numbers. If s_n converges to a number s, we say the series $\sum_{n=1}^{\infty} a_n$ converges and write $\sum_{n=1}^{\infty} a_n = s$. If an infinite series $\sum_{n=1}^{\infty} a_n$ converges means, the sequence of partial sums

$$s_1 = \sum_{n=1}^{1} a_n = a_1,$$

$$s_2 = \sum_{n=1}^{2} a_n = a_1 + a_2,$$

$$s_3 = \sum_{n=1}^{3} a_n = a_1 + a_2 + a_3,$$

A. S. R. Srinivasa Rao, *Mathematical Demography: Theory and Modeling*, Mathematical Marvels: Texts and Monographs in the Spirit of CR Rao,
https://doi.org/10.1007/978-981-95-6129-2

$$\vdots$$

$$s_m = \sum_{n=1}^{m} a_n = a_1 + a_2 + \cdots + a_m,$$

$$\vdots$$

converges. For example, if $a_n = a^n$ for $-1 < a < 1$, then the series $s_n = \sum_{n=0}^{\infty} a_n$ converges, because

$$s_n = \sum_{n=0}^{\infty} a_n = \sum_{n=0}^{\infty} a^n = 1 + a + a^2 + \cdots = \frac{1}{1-a} \quad (-1 < a < 1). \tag{E.1}$$

But s_n in (E.1) diverges if $a \geq 1$.

Let us define a sequence of functions $(s_n(x))_{n\geq 1}$ for each $x \in S$ as follows:

$$s_n(x) = \sum_{i=1}^{n} f_i(x), \tag{E.2}$$

where $\sum_{i=1}^{n} f_i(x)$ is called the series of functions $f_i(x)$ for $i = 1, 2, 3, \ldots n$. The series $\sum_{i=1}^{\infty} f_i(x)$ is called an infinite series. In (E.2), we will note the following sequence of functions:

$$\begin{aligned}
s_1(x) &= f_1(x),\\
s_2(x) &= f_1(x) + f_2(x),\\
s_3(x) &= f_1(x) + f_2(x) + f_3(x),\\
&\vdots\\
s_n(x) &= f_1(x) + f_2(x) + \cdots + f_n(x),\\
&\vdots
\end{aligned} \tag{E.3}$$

In (E.3), we call $s_n(x)$ the *nth* partial sum of functions. We say the series $\sum_{i=1}^{\infty} f_i(x)$ converges uniformly on S if there exists a function $f(x)$ such that $s_n(x)$ converges to f. Similar to the notation used to denote a convergent of series of numbers to a number, we write

$$\sum_{i=1}^{\infty} f_i(x) = f(x). \tag{E.4}$$

If

$$0 < |f_i(x)| < M_i$$

for $M_i \geq 0$, and $i = 1, 2, \ldots$ and for every x in S, then also we say the series $\sum_{i=1}^{\infty} f_i(x)$ converges uniformly on S.

Definition E.2 Weierstrass M-Test. Let $\{f_n\}$ be a sequence of real valued functions on S such that $|f_n(x)| \leq M_n$ for all $x \in S$. Given $\sum_n M_n < +\infty$, then $\sum_{i=1}^{\infty} f_i(x)$ converges uniformly for x in S.

Example E.3 The series $\sum_{n=1}^{\infty} \frac{1}{n^2} x^n$ converges uniformly for $-1 \leq x \leq 1$ with $M_n = \frac{1}{n^2}$.

Appendix F
Lebesgue Measure

First, we introduce the Riemann integral of a function f on a real-valued interval $[a, b]$. The Riemann interval is defined using two quantities, namely, *upper integral* and *lower integral*. These functions are defined using a partition of $[a, b]$. Let $\mathcal{X}$ be the collection of all possible partitions on $[a, b]$. A partition Y of $[a, b]$ is represented by $Y = \{a = x_0 < x_1 < \ldots < x_n = b\}$. Here $\mathcal{X}$ consists of all such Y on $[a, b]$. We define the upper integral of f, say $U(f)$ by $U(f) = \inf\{U(f, Y) : Y \in \mathcal{X}\}$, and the lower integral of f, say $L(f)$ by $L(f) = \sup\{L(f, Y) : Y \in \mathcal{X}\}$, where $U(f, Y) = \sum_{i=1}^{n} m_i(x_i, x_{i-1})$ for $m_i = \inf\{f(x) : x \in [x_i, x_{i-1}]\}$, and $L(f, Y) = \sum_{i=1}^{n} M_i(x_i, x_{i-1})$ for $M_i = \sup\{f(x) : x \in [x_i, x_{i-1}]\}$. Here, $U(f) \geq L(f)$ for any bounded function f on $[a, b]$. A bounded function f is Riemann integrable on $[a, b]$ if $U(f) = L(f)$. The quantities $U(f, Y)$ and $L(f, Y)$ are called upper and lower Riemann sums, respectively. Two immediate elementary consequences of Riemann integral are:

(i) If f is continuous on $[a, b]$, then f is integrable,

(ii) Let a real-valued function be defined on $[a, b]$, and let $\alpha \in (a, b)$. Then, f is integrable on $[a, b]$ if, and only if, f is integrable on both $[a, \alpha]$ and $[\alpha, b]$. That is,

$$\int_a^b f(x)dx = \int_a^\alpha f(x)dx + \int_\alpha^b f(x)dx.$$

The Riemann integral can also be defined on higher dimension.

Let S be a nonempty set in $\mathbb{R} = (-\infty, \infty)$, i.e., S has at least one element. A function $\varphi_S(x)$ is called the *characteristic function* if it is defined as follows:

$$\varphi_S(x) = \begin{cases} 1 & \text{if } x \in S \\ 0 & \text{if } x \in \mathbb{R} - S \end{cases}, \tag{F.1}$$

A. S. R. Srinivasa Rao, *Mathematical Demography: Theory and Modeling*, Mathematical Marvels: Texts and Monographs in the Spirit of CR Rao,
https://doi.org/10.1007/978-981-95-6129-2

where $\mathbb{R} - S = \{x : x \in \mathbb{R} \text{ and } x \notin S\}$. If $\varphi_S(x)$ is defined on an interval $I \subseteq \mathbb{R}$, then we call $\varphi_S(x)$ as a *measurable function*. A subset $S \subseteq \mathbb{R}$ is called a measurable set if $\varphi_S(x)$ is measurable. A measure of a set is a generalization of length of that set. Lebesgue integral is a generalization of standard Riemann integral. If two sets A and B are measurable, then $A \cup B$ and $A \cap B$ are measurable.

If $\varphi_S(x)$ is *Lebesgue integrable* on $\mathbb{R}$, and a measure μ of a set S, say $\mu(S)$, is defined by

$$\mu(S) = \int_{\mathbb{R}} \varphi_S(x)dx, \tag{F.2}$$

then μ is called *Lebesgue measure*. If $A \subseteq B$, then $\mu(A) \leq \mu(B)$. For further reading on measures and measurable functions, refer to [1, 6, 7].

Appendix G
Picard's Theorem and Differential Equations

Consider an IVP (initial value problem)

$$\frac{dy(t)}{dt} = f(t, y(t)), \quad y(t_0) = y_0.$$

Suppose $f(t, y(x))$ and $\frac{\partial f(t,y(t))}{\partial y}$ are continuous functions in some open rectangle R in a region D in the plane, where

$$R = \{(t, y) : a < t < b, c < y < d\}$$

for real numbers $\{a, b, c, d\}$, and it contains the point (t_0, y_0). Then, the IVP has a unique solution in some closed interval $I = [t_0 - \epsilon, t_0 + \epsilon]$ for $\epsilon > 0$. Here D is a region of set of points with the property that given (t_0, y_0), it is possible to construct a rectangle centered at (t_0, y_0) that lies in D. The function f is defined on D in the plane. See Fig. G.1. We will find an interval I on the $t-$axis and a function ϕ (which we call a solution of $\frac{dy(t)}{dt} = f(t, y(t)))$ such that $(i)\phi(t)$ and $\phi'(t)$ exist for each t in I and ϕ lies entirely in D, i.e., $(t, \phi(t))$ lie in D for all t, and (ii) for each t in I, we have $\frac{d\phi(t)}{dt} = f(t, \phi(t))$. In fact, ϕ is continuous of the triplet (t, t_0, y_0).

Theorem (Formal Statement) *Let $f(t, y(x))$ and $\frac{\partial f(t,y(t))}{\partial y}$ be continuous in a region D in the plane. Let (t_0, y_0)be a given point in D. Then there exists an interval containing t_0 and exactly one solution ϕ on I of $\frac{dy(t)}{dt} = f(t, y(t))$ passing through (t_0, y_0). The solution ϕ exists for those values of t for which the points $(t, \phi(t))$ lie in D.*

Let us consider a higher-order differential equation of the form

$$\frac{d^n y(t)}{dt^n} = f\left(t, y(t), \frac{dy(t)}{dt}, \frac{d^2 y(t)}{dt^2}, \ldots, \frac{d^{n-1} y(t)}{dt^{n-1}}\right).$$

A. S. R. Srinivasa Rao, *Mathematical Demography: Theory and Modeling*, Mathematical Marvels: Texts and Monographs in the Spirit of CR Rao,
https://doi.org/10.1007/978-981-95-6129-2

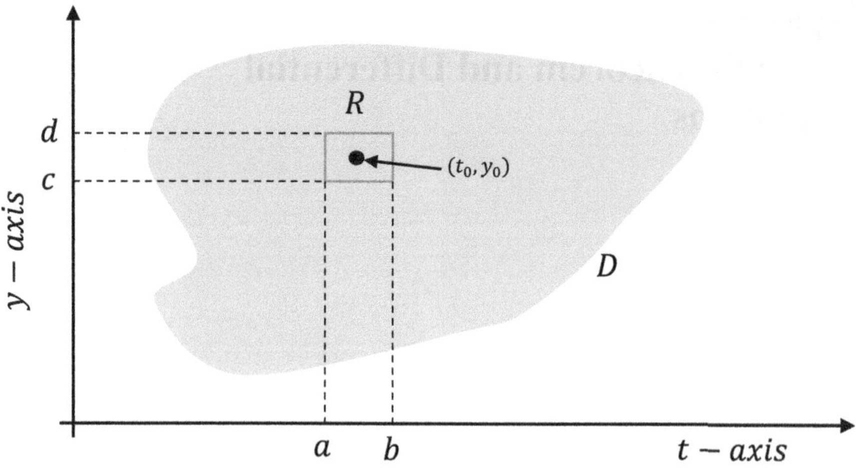

Fig. G.1 A point in a rectangle R contained in a region D

A function ϕ is a solution of this nth-order differential equation if $(i)\phi(t)$, $\frac{d\phi(t)}{dt}$, $\frac{d^2\phi(t)}{dt^2}$, and so on exists for each t in I, and the point

$$\left(t, \phi(t), \frac{d\phi(t)}{dt}, \frac{d^2\phi(t)}{dt^2}, \ldots, \frac{d^{n-1}\phi(t)}{dt^{n-1}}\right)$$

lies in D, (ii) for each t, the following holds:

$$\frac{d^n\phi(t)}{dt^n} = \left(t, \phi(t), \frac{d\phi(t)}{dt}, \frac{d^2\phi(t)}{dt^2}, \ldots, \frac{d^{n-1}\phi(t)}{dt^{n-1}}\right).$$

Picard also suggested an iteration formula for y_n represented by

$$y_{n+1}(t) = y_0 + \int_{t_0}^{t} f(s, y_n(s))\, ds$$

that generates a sequence of functions $(y_n(t))_{n\geq 1}$ that converges to the solution uniformly on I. We are not discussing here the proof of uniform convergence of $(y_n(t))$. The proof uses the Weierstrass M-test defined in Appendix E. The given IVP is written in integral form as follows:

$$y(t) = \int_{t_0}^{t} f(s, y(s))\, ds + y_0.$$

By considering the right side of the above integral as input, we get successive outputs as below:

$$y_1(t) = \int_{t_0}^{t} f(s, y_0(s))\, ds + y_0,$$

$$y_2(t) = \int_{t_0}^{t} f(s, y_1(s))\, ds + y_0,$$

$$\vdots$$

$$y_n(t) = \int_{t_0}^{t} f(s, y_{n-1}(s))\, ds + y_0.$$

By Picard's theorem, if $f(x, y(x))$ and $\frac{\partial f(x,y(x))}{\partial y}$ are continuous functions in R and $(t_0, y_0) \in R$, then there exists interval I containing t_0 and exactly one solution defined on I of this differential equation $\frac{dy(t)}{dt} = f(t, y(t))$ which passes through (t_0, y_0).

Appendix H
Population Growth Models

In this section, we derive exponential and logistic growth models of the population. Let us consider a model (exponential) of the form

$$\frac{dN}{dt} = rN, \tag{H.1}$$

where N is the size of a population, and r is the growth rate (or intrinsic population growth rate in the absence of migration data). The quantity r is also expressed as $b - d$, where b is the birth rate and d is the death rate of the population. We will solve for $N(t)$, i.e., the size of the population at time t using the model in (H.1). This model can be written as

$$\begin{aligned}
\frac{1}{N}\frac{dN}{dt} = r \implies & \int \frac{1}{N} dN = \int r dt = rt + C \\
\implies & \log N(t) = rt + C \text{ (C is the constant of integration)} \\
\implies & N(t) = \exp(rt + C) = \exp(rt)\exp(C) = N(0)\exp(rt),
\end{aligned} \tag{H.2}$$

where $N(0) = \exp(C)$. When $r < 0$ in (H.2) the population decays, and when $r > 0$ in (H.2), the population grows. When $r = 0$ in (H.1), then $\frac{dN}{dt} = 0$ indicating no growth in the population. A following revised model for (H.1) is written using an additional parameter K referred to as the saturation point of population growth:

$$\frac{dN}{dt} = rN\left(1 - \frac{N}{K}\right). \tag{H.3}$$

Equation (H.3) is often referred to as a logistic growth population model.
See Fig. H.1 for exponential model and Fig. H.2 for logistic model. Let us now obtain the solution $N(t)$ for the logistic model. Rearranging the model in (H.3) and

A. S. R. Srinivasa Rao, *Mathematical Demography: Theory and Modeling*,
Mathematical Marvels: Texts and Monographs in the Spirit of CR Rao,
https://doi.org/10.1007/978-981-95-6129-2

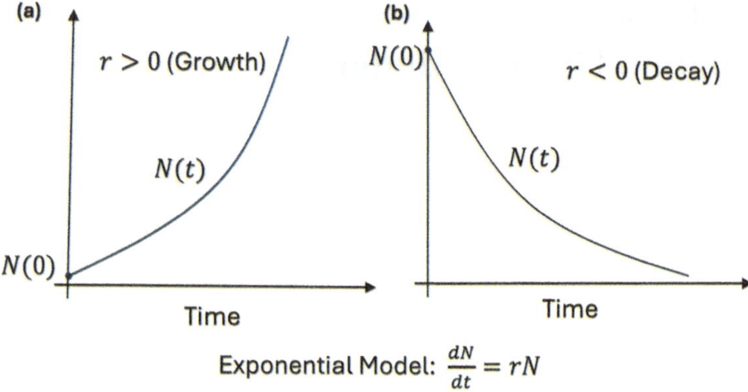

Fig. H.1 Exponential model. (**a**) Population grows when $r > 0$ and (**b**) population decays when $r < 0$

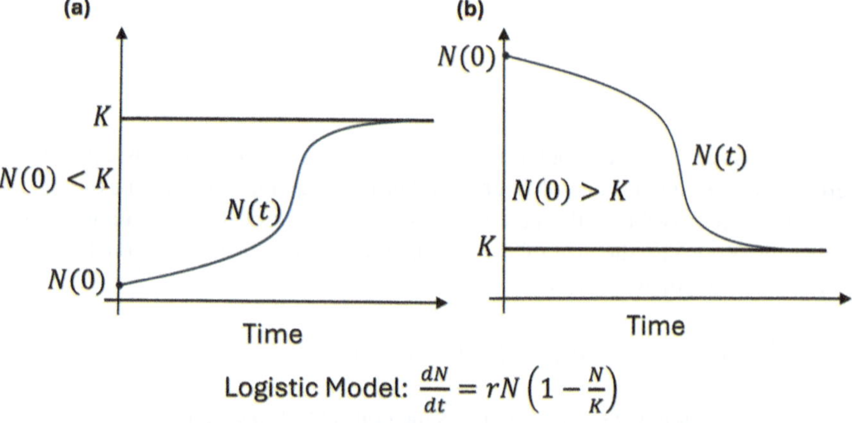

Fig. H.2 Logistic model. (**a**) Initial population $N(0) < K$ and (**b**) initial population $N(0) > K$. Here K is the carrying capacity

integrating, we get

$$\int \frac{KdN}{N(K-N)} = \frac{r}{k}\int dt$$

$$\Longrightarrow \frac{1}{K}\int\left(\frac{1}{N}+\frac{1}{k-N}\right)dN = \frac{r}{K}t + C_1 \text{ (C_1 is the constant of integration)}$$

$$\Longrightarrow \frac{1}{K}\left[\log N - \log(K-N)\right] = \frac{r}{K}t + C_1$$

$$\Longrightarrow \frac{N}{K-N} = \exp\left(\frac{r}{K}t\right)\exp\left(C_1\right) = C_2\exp\left(\frac{r}{K}t\right), \tag{H.4}$$

where $C_2 = \exp(C_1)$. This gives us

$$N(t) = \frac{KC_2 \exp(rt)}{1 + C_2 \exp(rt)}. \tag{H.5}$$

Dividing the numerator and denominator of the right side of (H.5) by $C_2 \exp(rt)$, we get

$$N(t) = \frac{K}{\frac{1}{C_2 \exp(rt)} + 1} \tag{H.6}$$

$$= \frac{K}{1 + N(0) \exp(-rt)} \text{ (where } \frac{1}{C_2} = N(0)). \tag{H.7}$$

One can consider either of the forms expressed in (H.6) and (H.7).

Appendix I
Jacobian Matrix, Eigenvalues, Equilibrium Points

Let $x \in \mathbb{R}^n$, where $\mathbb{R}^n$ is an Euclidean space, i.e., the space of all $n-$tuples of real numbers. That is, $x = (x_1, x_2, \ldots, x_n)$. Let $y \in \mathbb{R}^n$, and then $x + y$ is defined as $x + y = \{x_1 + y_1, x_2 + y_2, \ldots, x_n + y_n\}$. For a scalar c, we define $cx = \{cx_1, cx_2, \ldots, cx_n\}$. We also denote $\mathbb{R}^n$ for $n-$component column vectors of real numbers, i.e., for $x \in \mathbb{R}^n$, we can write $x = \begin{pmatrix} x_1 \\ \vdots \\ x_n \end{pmatrix}$. Let $f : S \to \mathbb{R}^n$ be a function for an open set $S \in \mathbb{R}^n$. Here f is a multivariate function with variables $x_1,\ldots, x_n$, and we can write $f(x) = \begin{pmatrix} f_1(x) \\ \vdots \\ f_n(x) \end{pmatrix}$. If $f'(x)$ exists for $x \in S$, then $\frac{\partial f_i(x)}{\partial x_j}$ exists for all i and j, and the determinant of the linear operator $f'(x)$ is called the Jacobian of f at x. The Jacobian is expressed as

$$J_f(x) = \det f'(x). \tag{I.1}$$

The Jacobian is also expressed as

$$\frac{\partial f(x_1, \ldots, x_n)}{\partial(x_1, \ldots, x_n)}. \tag{I.2}$$

If the derivative of f exists at some x in S, then it has to be the linear function whose matrix is $\left[\frac{\partial f_i(x)}{\partial x_j}\right]$. This matrix is called the Jacobian matrix of f at x. If each $\frac{\partial f_i(x)}{\partial x_j}$

A. S. R. Srinivasa Rao, *Mathematical Demography: Theory and Modeling*, Mathematical Marvels: Texts and Monographs in the Spirit of CR Rao,
https://doi.org/10.1007/978-981-95-6129-2

exists at every point of S and is continuous at a point x_0, then $f'(x_0)$ exists. We define

$$\frac{\partial^2 f_i}{\partial x_i \partial x_j} = \frac{\partial}{\partial x_i}\left(\frac{\partial f_i}{\partial x_j}\right). \tag{I.3}$$

Now, let us consider S in $\mathbb{R}^2$ and f is defined in $\mathbb{R}$, i.e., $f : S \to \mathbb{R}^2$. Let (x, y) be a point in S. If $\frac{\partial f}{\partial x}$, $\frac{\partial f}{\partial y}$, and $\frac{\partial^2 f}{\partial x \partial y}$ exist in S, and $\frac{\partial^2 f}{\partial x \partial y}$ is continuous at $(x, y) = (x_0, y_0)$, then $\frac{\partial^2 f}{\partial x \partial y}(x_0, y_0)$ exists, and $\frac{\partial^2 f}{\partial x \partial y}(x_0, y_0) = \frac{\partial^2 f}{\partial y \partial x}(x_0, y_0)$. When $x = r\cos\theta$ and $y = r\sin\theta$, the Jacobian matrix J is obtained as

$$J = \begin{pmatrix} \frac{\partial x}{\partial r} & \frac{\partial y}{\partial r} \\ \frac{\partial x}{\partial \theta} & \frac{\partial y}{\partial \theta} \end{pmatrix} = \begin{pmatrix} \cos\theta & \sin\theta \\ -r\sin\theta & r\cos\theta \end{pmatrix}. \tag{I.4}$$

A Hessian matrix on a function (H_f) which is a second-order partial derivative matrix combined with J gives a Taylor's series expansion of multivariate function. The $(H_f)_{i,j}$ is defined as $(H_f)_{i,j} = \frac{\partial^2 f_i}{\partial x_i \partial x_j}$. A three-dimensional H_f is expressed as $H_{ijk} = \frac{\partial^2 f_i(x)}{\partial x_i \partial x_j}$, where $x = (x_1, x_2, x_3)$. The Taylor expansion of $f(x)$ around a point a is expressed as

$$f_i(x) \approx f_i(a) + \sum_j J_{ij}\big|_a (x_j - a_j) + \frac{1}{2}\sum_j \sum_k H_{ijk}\big|_a (x_j - a_j)(x_k - a_k). \tag{I.5}$$

Let A be a matrix and x be a nonzero vector. For a scalar λ, if we solve $Ax = \lambda x$ for λ, then the values of λ are called the eigenvalues and x is called the eigenvector of A. If A is a real-valued matrix and λ is a real eigenvalue of A, then $(\lambda I - A)x = 0$ (for identity matrix I) has a nontrivial solution over $\mathbb{R}$. There could be non-real eigenvalues of A as well (a situation we saw in stability analysis of Lotka–Volterra two-population model). The equation $\det(Ax - \lambda I) = 0$ is called the characteristic equation of A, and the reason λs are also referred as characteristic roots of A. A set of eigenvectors $x_1, x_2, \ldots, x_n$ are linearly independent if

$$\lambda_1 x_1 + \lambda_2 x_2 + \ldots + \lambda_n x_n = 0 \implies \lambda_1 = \lambda_2 = \ldots = \lambda_n = 0; \tag{I.6}$$

otherwise, if $\lambda_1, \lambda_2, \ldots \lambda_n$, not all zero, with the condition that $\lambda_1 x_1 + \lambda_2 x_2 + \ldots + \lambda_n x_n = 0$, then the eigenvectors are linearly dependent. Let $\lambda_1, \lambda_2, \ldots \lambda_n$ be distinct eigenvalues of A and $x_1, x_2, \ldots, x_n$ be corresponding eigenvectors; then these eigenvectors are linearly independent. Let $\lambda_1, \lambda_2, \ldots \lambda_n$ be the eigenvalues of A and let $f(\lambda)$ be a polynomial; then, $f(\lambda_1), f(\lambda_2), \ldots, f(\lambda_n)$ are the characteristic roots of $f(A)$.

Consider a point $x^* \in \mathbb{R}^n$. The point x^* is called the equilibrium point for the differential equation $\frac{dx}{dt} - f(x, t)$ if $f(x^*, t) = 0$ for all t. An equilibrium point is also called a critical point. These equilibrium points combined with the eigenvalues help us to understand whether a differential equation system is stable or unstable. We also briefly discussed these in Chap. 4.

References

1. Apostol, T.M.: Mathematical Analysis, 2/e. Addisson-Wesley, New York and Narosa Publishers (India), New Delhi (1973)
2. Shirali, S., Vasudeva, H.L.: Metric Spaces, viii+222 pp. Springer, London (2006)
3. Churchill, R.V., Brown, J.W.: Complex Variables and Applications, 4th edn., x+339 pp. McGraw-Hill Book Co., New York (1984)
4. Krantz, S.G.: Complex Analysis: The Geometric Viewpoint, 2nd edn. Carus Mathematical Monographs, vol. 23, xviii+219 pp. Mathematical Association of America, Washington, DC (2004)
5. Rao, A.S.R.S.: Multilevel contours on complex planes. School Sci. **61**(2&3), 5–8 (2023)
6. Rao, C.R.: Linear Statistical Inference and Its Applications, 2nd edn. Reprint. Wiley India edition. Wiley, New Delhi (2006)
7. Abbott, S.: Understanding analysis. Undergraduate Texts in Mathematics, xii+257 pp. Springer, New York (2001)
8. Kainth, S.P.S.: A Comprehensive Textbook on Metric Spaces. Springer (2023)

Index

A. S. R. Srinivasa Rao, *Mathematical Demography: Theory and Modeling*,
Mathematical Marvels: Texts and Monographs in the Spirit of CR Rao,
https://doi.org/10.1007/978-981-95-6129-2

The manufacturer's authorised representative in the EU is Springer Nature Customer Service Centre GmbH, Europaplatz 3, 69115 Heidelberg, Germany. If you have any concerns regarding our products, please contact ProductSafety@springernature.com

Printed and bound by CPI Group (UK) Ltd, Croydon, CR0 4YY
23/07/2026
02175610-0001